Insider's Guide to Quantum Physics Exposed

Kalum .K Greenwood

All rights reserved.

Copyright © 2024 Kalum .K Greenwood

Insider's Guide to Quantum Physics Exposed : Unpacking the Mysteries of Quantum Physics: Essential Insights for All Readers.

Funny helpful tips:

Stay connected with a fitness community; shared goals and support can be motivating.

Collaborate and partner; synergies can lead to expanded reach and resources.

<u>Life advices:</u>

In the mountains of challenges, scale new heights with determination and grit.

Avoid holding onto past mistakes; focus on the present and future.

Introduction

Embark on a journey through the fascinating world of quantum physics in this beginner's guide, where you'll explore the history, key principles, and significant advancements in this groundbreaking field.

The guide begins by delving into the infancy of quantum physics, highlighting the early atomic models and pivotal experiments like the double-slit experiment. It introduces the concept of cathode rays, black-body radiation, and the groundbreaking discovery of radioactivity.

As the early 20th century witnessed a profound transformation in physics, the guide unravels the parting of ways between classical and quantum physics.

Waves, particles, and sub-atomic entities take center stage as the guide navigates through the quantum realm's inherent uncertainty. Schrödinger and his groundbreaking wave mechanics, including the famous Schrödinger's equation, are presented in an accessible manner.

The guide also sheds light on the genius of Albert Einstein, exploring his contributions to physics, such as the theory of relativity and the photoelectric effect. Einstein's journey, both as a scientist and a person fleeing political turmoil, is woven into the narrative.

Advancements in quantum physics during the mid-20th century are examined, from the works of Paul Dirac to nuclear developments in times of war and peace. The guide touches on how physics has been harnessed for healing and the construction of massive scientific instruments.

Observing the tiniest objects and the farthest reaches of the universe is a pivotal aspect of quantum physics, and this guide explains the methods and technologies involved. It also looks forward to the 21st century, where everyday objects can perform extraordinary tasks thanks to quantum principles.

As you journey through the pages of this book, you'll gain a foundational understanding of this complex yet captivating field, appreciating its historical significance and its potential to shape the future of science and technology.

Contents

Chapter 1: The Infancy of Quantum Physics 1
    Early Atomic Models 2
    The Double-Slit Experiment 5
    Cathode Rays and Black-Body Radiation 8
    Radioactivity- What a Killer Discovery! 9

Chapter 2: Early 20th-Century Physics-Building a Whole New Discipline 13
    The Parting of Ways Between Classical and Quantum 13
    The BIG Principles You Need to Know 14
    Measuring Large Numbers of Tiny Things 17
    Planck and His Constant 18
    Origin of the Theory of the Photoelectric Effect 20
    Waves, Particles, and Sub-atomic Stuff 23
    So Much Uncertainty 26

Chapter 3: Schrödinger, His Theories, and His Cats 28
    Schrödinger's Quantum Theory and Unified Field Theory 28
    Congratulations! It's Wave Mechanics 29
    Schrödinger's Equation and Atomic Model 30
    Yes, Yes, the Cats 31

Chapter 4: It's All Relative - The Genius of Einstein 34
    A Portrait of the Scientist as a Young Man 34
    Miracle Year: Unlocking the Secrets of the Photoelectric Effect 36
    Special Relativity, Mass Equivalency, and General Relativity 38
    Leaving Germany Behind for Good 46

Chapter 5: Even Einstein Can't Always Be Right 49

    Debating the State of Physics 49
    Outside the Classroom 55
    Back to Science, Sort Of 58
Chapter 6: Rapid Advances in the Mid-20th Century 61
    Don't Dirac the Boat 61
    Picking Back Up with Particle Theory 64
    Nuclear Advancements in War and Peace 68
    Harnessing Physics for Healing 71
Chapter 7: Building Big Things…forScience! 74
    Observing the Tiniest Objects 74
    Observing the Farthest Objects 76
    Manipulating the Atom 77
Chapter 8: What's Next? Quantum Physics in the 21st Century and Beyond 80
    Everyday Objects Doing Extraordinary Things 80
    Got the World on a String 82

Chapter 1: The Infancy of Quantum Physics

Scientists didn't just wake up one day with a laboratory full of fancy electron-scanning microscopes and mass spectrometers. It took years of discovery and experimentation to get to the understanding of quantum physics that we have today. The story of quantum physics, which holds a dictionary definition of "the branch of physics concerned with the quantum theory" (helpful, huh?), begins with the story of classical physics. This, of course, is the scientific discipline that aims to use mathematics and mechanics to explain how the world works. So, where does the fundamental split between classical and quantum physics occur?

The truth is, there is no specific date or definitive line that shows us where classical physics ends and quantum theory begins. Like a snowball rolling downhill, the theories that would evolve into quantum studies started slowly, building up momentum and growing- and that snowball eventually became an avalanche that is still roaring today. It was innate human curiosity that instigated the first experiments that would lead to the creation of quantum physics as a separate field of study. When scientists found that they couldn't explain things using the laws of classical physics, they went looking for their answers elsewhere. Thus, quantum physics, quantum mechanics, quantum chemistry, and every other quantum discipline was born.

Even before the apple could fall on Newton's head and spark the origin of classical physics, quantum physics made a brief debut. The earliest concept of the atom was actually introduced circa 400 BCE by the Greek philosopher Democritus. He proposed a so-called Theory of the Universe that subscribed to the following principles:

- all matter consists of atoms; these are too small to be seen
- there is empty space between all atoms
- atoms are solid but have no set internal structure
- every substance is made of different atoms, with their own weight, size, and shape

This early theory is surprisingly accurate, albeit generalized. Given that Democritus based his theories on philosophical thought experiments rather than scientific study, this theory's accuracy is even more surprising. He firmly believed that all substances could be divided and divided again and again until only the smallest elements of that matter remained- and that smallest particle is that which he dubbed "the atom." His theory would remain widely accepted but largely untouched until the turn of the 19th century.

Early Atomic Models

Sometime shortly after 1800, British scientist John Dalton turned his attention away from his eclectic studies in meteorology and physiology and tried to figure out Democritus's atomic theory, making him the first modern researcher to put any serious thought into the centuries-old hypothesis. Dalton released his atomic theory in 1808, which rehashed many of Democritus's ideas but also threw in a new twist, the introduction of atomic weights to determine one substance from another. Dalton's Atomic Theory looked like this:

- all matter is made up of tiny particles known as atoms
- atoms cannot be destroyed or changed
- atoms and elements are distinguishable by their weights
- elements react to each other and form new compounds out of their combined atoms

While Dalton's model was closer to the truth and more accurate than Democritus's, it wasn't without controversy. Unfortunately, the controversy stemmed not from the content of the hypothesis, but from an accusation by Irish chemist William Higgins that Dalton had stolen his work. It turns out Dalton hadn't. He'd actually taken much of his research from Higgins's uncle, Bryan Higgins, a naturalist and chemist himself. That's some defense, isn't it? Imagine Dalton pounding his fist on a table, hollering, "How dare you accuse me of stealing your ideas?!? I would never! I stole your uncle's, as they are much better!"

In all factuality, Dalton did base his concepts on both the original atomic theory proposed by Democritus and the elder Higgins, but the theory he released was proven to be very much his own. Notebooks from Dalton's laboratory prove that he was engaged in unique, proprietary experiments that even went so far as to suggest the relative weights of several elements. Despite not having many original ideas, he did study the concepts in a new way, and for that, he is the great-grandfather of the atomic model. Dalton's papers were preserved and archived in Manchester, England, but many were lost in the 1940 Christmas Eve blitzkrieg.

The next atomic theory was proposed by British physicist JJ Thomson, and it bears the fun moniker "the plum pudding model." Thomson is widely credited with discovering the electron using cathode rays (more on those shortly!) and consequently released his revamped atomic model in 1904. Plum pudding, for those not well-versed in British desserts, is a sweetened bread pudding chock full of butter and brown sugar and flocked with raisins or currents. It was, and still is, a favorite in the United Kingdom. With his discovery of the electron, Thomson actually took the atomic model farther away from accuracy. He had discovered *what* they were (negatively charged parts of the atom) but not *how* they worked.

Thomson didn't do much to change the atomic theory of the past century, but his model was, frankly, a little wonky. Thomson's plum pudding model showed a field of positively charged...goo? Atomic gel?...no one's quite sure... that was sprinkled with negatively-charged electrons. This resemblance to the tasty treat is what lent the atomic model its name. Despite his model's inaccuracy, Thomson's contribution to the discovery of the electron shouldn't be discounted- it was undoubtedly a significant advancement in the fledgling field of quantum physics.

Dissatisfied with Thomson's model, physicist Ernest Rutherford was the next to tackle atomic theory. Rutherford is often called the grandfather of nuclear physics and was a colleague of many of the other brilliant scientific minds of the turn of the 20th century, including JJ Thomson. Something about Thomson's atomic model just didn't sit right with Rutherford. So he set out to determine what exactly was amiss with it. The answers came while Rutherford was supervising the now-famous Geiger-Marsden gold foil experiments at the University of Manchester in England.

This groundbreaking research involved bombarding a thin piece of gold sheeting with a stream of alpha particles (ionized helium) and observing the reaction of the atoms in the alpha stream. To the men's surprise, they could discern that the alpha particles were not all passing through the gold foil, but some were instead being deflected away from the foil. Why was this so important, and how did it lead Rutherford to create a new model of the atom?

To Rutherford, this observation signaled that the atom could not be a gelatinous blob as the Thomson plum pudding model portrayed. Rutherford believed that this must mean that the helium atoms had a hard center of some sort that wouldn't pass through the gold. He was right and had inadvertently discovered that atoms have a nucleus. Thus, in 1911, after several years of experimentation, the plum pudding model was relegated to the scientific trash can, and the Rutherford model was introduced to the world. Rutherford's atom was made up of a nucleus of positively-charged protons, held closely together enough to not be able to always pass through another substance, and orbits of negatively-charged electrons.

This model was in vogue for two short years, only to be replaced by the Rutherford-Bohr model, or simply the Bohr model. Sorry, Rutherford! (But don't feel too bad for him, he won the Nobel Prize, was knighted, and was given a barony for his scientific achievements.) The Bohr model was the first to show electrons in layered orbits around the nucleus, leading scientists even closer to creating an accurate picture of the tiny atom. This model is still taught to younger students as a clear, easy-to-decipher depiction, although it is not entirely correct.

The atomic model that is widely accepted today is the one created by Erwin Schrödinger (so much more on this gem of a scientist later!) in 1926. Schrödinger is credited with the design of the atomic cloud, which shows the electrons moving not in set orbits like the Bohr model, but in waves fluctuating in and out from the nucleus. This model was based on the theory of wave-particle duality, which we'll talk about momentarily. For now, let's review the evolution of atomic models- Democritus, Dalton, Thomson, Rutherford, Bohr, Schrödinger- and if you can come up with a mnemonic device to remember that, good on you!

The Double-Slit Experiment

One thing that really jump-started the move from classical physics into quantum theory is the double-slit experiment, which became the origin of wave-particle duality. The double-slit experiment's basic premise is to send a beam of concentrated energy- like light or laser- at a solid surface with two narrow slits cut into it, projecting the energy through the slits and onto a flat surface behind it. Why would we do this, and what does it prove?

The double-slit experiment results in a phenomenon that shows light does not behave as either a wave or a particle but acts as both. This is wave-particle duality at its most basic. This discovery is an important foundational principle of quantum theory. When the light source is shone through a single slit, it shows up on the rear surface as a solid bar of illumination. When it is shone through a pair of slits, the light is broken up and displays on the rear surface in a striated pattern. The reason this occurs is that the light is acting in accordance with wave-particle duality.

When the double-slit experiment was first conducted in 1801, it was believed that light could only behave as a wave or a particle. Once it was determined that it could be both, serious work was undertaken on the duality theory. Until 1927, scientists thought that only light could behave this way, but it was proven that electrons also have wave-particle duality. Later on, it was shown that all atomic particles have wave-particle duality.

So what exactly IS wave-particle duality, and why is it so important to the study of quantum physics? For starters, let's define a wave in terms of science. A wave is a disturbance in a medium that facilitates the movement of energy. This could be sound waves through air or water, radio waves through the atmosphere, or kinetic energy waves like ripples from a stone dropped in a puddle. Waves have an amplitude (the height of the wave) and frequency (the speed at which they cycle through an entire wavelength). These characteristics define what type of energy the wave is conveying.

Physics, both classical and quantum, focuses on the waves that make up the electromagnetic spectrum. The spectrum wasn't discovered all at once by the same researchers but was more of a jigsaw puzzle put together piece by piece until the big picture could be seen. Between 1800 and 1915, each discovery lent a new understanding of the electromagnetic spectrum until it became the scale that is still used today. The scale begins on the left with the highest frequency and slides through the spectrum to the lower frequency on the right like so:

Gamma rays
--
X-rays
--
Ultraviolet light
--
Visible light (violet, indigo, blue, green, yellow, orange, red)
--
Infrared light
--
Microwaves
--
Radio waves
--

The completion of the electromagnetic spectrum still left scientists with a few questions. If waves needed a medium to travel through, why could we see the light from the Sun through the vacuum of space? It turns out that electromagnetic waves are pretty unique- they don't *need* a medium to travel through. This was a startling revelation that began to set quantum physics apart from classical physics. But it still left researchers scratching their heads because if light and other electromagnetic waves didn't need to be moved through a medium, then were they really waves, or were they self-propelling particles? Well, both. Sort of. Yeah.

You can try to wrap your head around this, but if you can't, don't lose sleep over it. There are people who get paid to think about this, and even they couldn't fully explain it. Just know that quantum theory shows us that the tiniest parts of the universe can act as both waves and particles, and it's pretty freaking cool. That's all you need to remember.

Cathode Rays and Black-Body Radiation

Let's talk about some sci-fi sounding stuff for a little bit because when these things started being used in laboratories, they truly were the product of imagination. JJ Thomson of plum pudding atom fame didn't just wake up one morning and find electrons dancing through his bedroom- he had to work out a scientific method to discover them and prove their existence, and he did so with cathode rays.

A cathode-ray tube is a neat little piece of scientific equipment that consists of a glass tube, which is evacuated of its inner gases before being sealed off. The result is a vacuum or near-vacuum environment. At one end, a metal cathode is electrified with a high-powered charge that is channeled through an anode into a concentrated beam or cathode ray. Two metal plates with magnets are placed on the outside of the tube, one positively charged and the other negatively charged. As the cathode ray travels down the tube, scientists are able to measure the movement and charge of the atoms and particles inside the tube. It was while conducting a cathode ray experiment that Thomson first stumbled upon the existence of electrons. He realized that the atoms inside the tube were not behaving in the manner he expected them to. Upon further trials, Thomson came to the conclusion that there must be something causing a fluctuation in the weight and action of the particles he was trying to measure.

Another major sci-fi-esque advancement in the infancy of quantum physics was the discovery of black body radiation. It sounds so spooky and mysterious, but once it's explained, it really helps make a lot of sense of the parts of the universe we can't readily see. Black body radiation is also called thermal radiation, and what it tells us is how energy equilibrium is maintained throughout all known matter. One of the best examples of an unbalanced, total measure of black body radiation is a black hole. It absorbs all energy that comes near it- no matter the type of wavelength.

More frequently, black body radiation helps scientists identify matter which cannot be seen by measuring the thermal activity of the waves in the area. Since a black body is defined as a body of matter which absorbs all electromagnetic energy that comes its way and then radiates it back out according to the laws of thermodynamics, think about a sidewalk or parking lot that gets burning hot on a summer day because it's absorbed the heat and light radiation from the Sun. At night, that pavement will re-radiate that energy back out into the air in an attempt to reach equilibrium. This is what black body radiation should also do, black holes and their enormous appetites aside. By tracking the movement of electromagnetic waves and seeing where they disappear and then reappear, scientists can hypothesize about the size and scope of black bodies within their laboratories and in the vastness of outer space.

Radioactivity- What a Killer Discovery!

Today, we know that radioactive elements and radioactive waves have a number of useful applications, but we also know that certain radioactive materials have an ill-effect on live tissue. This is why lead shields are used as protection during x-rays and why no one will ever live in Chernobyl again, at least not for the foreseeable future. In the late 1800s, William Roentgen discovered that x-rays could be used to "see inside" the human body, and his colleague, third-generation French scientist Henri Becquerel, was thoroughly intrigued by the find. Becquerel wondered if x-rays could explain a phenomenon he saw in his own work with fluorescent and phosphorescent elements and minerals.

Becquerel was also a skilled photographer and came up with the brilliant plan of trying to determine if his phosphorescent minerals were absorbing light and emitting it back in the form of x-rays. The experiment seemed simple enough- expose the elements to light, then tuck them between two photographic plates wrapped in black paper. If the elements were emitting x-rays, they should expose onto the plates like a photograph, right? Great theory, and at first, it worked. Becquerel repeated the experiment several times outside on bright, sunny days and was able to see the outlines of the uranium salt crystals on the photo plates. He also threw in some solid objects like coins for control purposes. Becquerel wanted to make sure he wasn't just capturing images of random dust particles and such. The scientist was confident that he'd found evidence that his elements were emitting x-rays that he wrote and submitted papers about it to some of his scientific academies.

Becquerel continued to play around with trying to get images of the x-rays themselves, until one day it was cloudy, and he couldn't achieve the lighting effect that he wanted. So he did what any good researcher would do- he tossed his materials in a drawer and went about his other business. A few days later, on March 1, 1896, he opened the drawer, and for some reason (perhaps the innate drive of scientific curiosity, or maybe just the need to have something to report at his next academy meeting?) Becquerel decided to develop the photo plates, despite them not being exposed to sunlight.

And what he got were the clearest pictures yet! How could that happen if the uranium salts hadn't been exposed to sunlight before being put between the plates? The only clear conclusion was that the material was producing its own form of radiation, one not dependent on being subject to light. Becquerel was astounded. He repeated the accidental experiment several times, trying out different types of uranium compounds to determine if the photo exposures were from leftover phosphorescence or from x-rays. But no matter how many variations he tried, Becquerel came up with the same results. He had accidentally discovered radioactivity by shoving his stuff into a drawer.

Fellow Frenchman Pierre Curie and his Polish-born wife Marie were conducting their own experiments with uranium, and they were fascinated by Becquerel's work. It was actually Marie who coined the term "radioactivity," and it was Marie who would pay the heaviest price for their work with uranium and the other elements that she discovered, polonium (named for her homeland) and radium (named for its high radioactivity). Radioactivity has given the world a lot of wonderful things- nuclear power, the ability to date ancient objects, alarm clocks that glow in the dark- but it has also caused a lot of tragedy.

That's because radioactivity, the ability of atoms to produce their own wave radiation as they naturally break down, is a phenomenon based on instability. Becquerel and the Curies would earn a Nobel Prize for their early work with radioactive elements, and Marie Curie would later become the first woman to win two Nobel Prizes, but she paid *with* her life *for* her life's work. Pierre Curie had his own life cut short by a road accident in 1906, but Marie soldiered on. She would create mobile x-ray units for ambulances during World War I, developed detailed theories and kept records on the radioactive decay of several elements, and raised her and Pierre's two daughters alone as a working scientist. Marie Curie developed blood cancer as a result of her overexposure to radioactivity but continued to drive for scientific advancement until her death in 1934. She is buried in a lead-lined casket to contain the radiation that is predicted to be emitted from her remains for at least 1200-1500 years.

What can the accidental discovery of radioactivity and the work of Becquerel (who also died from complications of radiation exposure) and the Curies tell us about the nature of research? To being with, it takes away some of the shine and stuffiness of academia. If Becquerel hadn't shoved his experimental items in a desk drawer, it might have taken years before radioactivity was pinpointed. It also tells us a lot about the uncharted territory that early quantum physicists and chemists were wading in. Two of the three scientists who were lauded for their work with radioactive elements died from the very act of doing so. And the third- well, that's just tragic. Who knows what else Pierre Curie would have contributed to early quantum studies?

If it seems like it's tough to keep every development in the infancy of quantum physics in chronological order, that's because it is, unfortunately. These discoveries were cascading so quickly, and experiments were being performed by such a broad set of laboratories and scientists that it's difficult to keep tabs on them all. We promise it will all start to become clearer as we move forward. In the next chapter, we'll take a look at where all these early discoveries were headed and get into some of the math and science behind what are now the foundational principles of quantum studies. Shall we?

Chapter 2: Early 20th-Century Physics- Building a Whole New Discipline

There was so much happening in the scientific world from the late 1800s through the early decades of the 1900s that it's almost impossible to keep up with every single development in chronological order. In the first chapter, we covered an overview of some of the big concepts without going into too much technical detail. Moving forward, we're going to start looking at some of the foundational principles of quantum physics beyond the model of the atom and some pretty cool early discoveries. We'll also (sorry!) have to look at some of the crazy math that quantum physics relies on to explain the behavior of matter and energy.

The Parting of Ways Between Classical and Quantum

As we said previously, there is no true break-up date between classical physics and quantum physics, but it became apparent early on that there would have to be a divorce. Classical physics deals with things that are causal. A classical physicist can measure and predict outcomes based on past results, repeat performance, and variables that have stood the test of time.

Quantum physicists handle data that behaves erratically, cannot often be measured without the use of mathematics rather than observations, and often, wave-particle duality and statistical probability play a role. Classical physics can tell you how long it will take a can of corn to roll across a parking lot when it falls through your shopping bag. Quantum physics can tell you how many molecules of gas and liquid are inside the can and how fast those molecules are vibrating as the can rolls.

The truth is, to understand the universe, you have to recognize that there are certain boundaries that have to exist. We can travel faster than sound, but we can't travel faster than light. Sadly, we can't time-travel, at least not in any significant way (see the theory of relativity later in this book.) Humans are capable of many things, but we aren't yet capable of breaking the laws of physics. Classical physics doesn't often attempt to stretch these boundaries. That discipline is focused on explaining the past to predict the future and the present to surmise the past. Classical physics also treats waves and particles as two separate concepts, and because of this, it has two sets of equations and principles to deal with them. Quantum physics recognizes that wave-particle duality exists and has adapted its mathematics to handle this reality. As we said, though, there are some ground rules you should know, and these hold true for both classical and quantum physics.

The BIG Principles You Need to Know

There are a few things you should always remember when trying to wrap your head around physics of any kind, and these are the law of conservation of matter, the law of conservation of energy, and the laws of thermodynamics. As vast as the universe is, we only have a finite, defined amount of, well, everything. These laws explain to us how that works, exactly.

The law of conservation of matter (sometimes called the law of conservation of mass) tells us a few things. First, it tells us its most important statement: matter cannot be created or destroyed. What we have is what we have, no more, no less. You cannot create something out of nothing. Secondly, the law of conservation of matter tells us that matter can change form, but the total mass of the new substance must equal the mass of the old substances. This is how and why chemical compounds work. Water (H_2O) is made up of two hydrogen atoms and one oxygen atom. The mass of the water molecule is equal to the mass of the hydrogen atoms and the oxygen atom. No mass has been lost in the creation of the water molecule. By using this law, we can calculate the mass in a number of different theoretical situations. The difference in mass should always be net-zero, assuming, of course, that you have a closed system. Introducing an unknown variable might throw your calculations off.

The other law that anyone who wants to understand physics needs to know is the law of conservation of energy. Like the law of conservation of matter, this law states that energy cannot be created or destroyed. Energy is also a finite resource. Every object in the universe, from the tiniest subatomic particle to the largest, brightest star, has a certain amount of energy that it is allotted. Energy can be transferred and transformed, but its quantity cannot be altered. But what about a resting object, you say? Surely it's not using any energy. Nope, it's not using it, but it is storing it as potential energy. Once it begins to move, according to classical physics, it will be expending that potential energy as kinetic energy. Quantum physics says that all matter is always moving, even if it is terribly slow and nearly indetectable. This is one of the principal schisms between classical and quantum physics.

Related to the law of conservation of energy is the law of thermodynamics. We see this in action every day inside our own homes. Heat is a form of energy; it is what happens when that energy is transferred between things that are not the same temperature. If you open a can of soup, and if you're as classy as we are, you put it into a bowl and pop it into the microwave. The energy from the microwaves causes the molecules in your soup to move faster, bump into each other, cause friction, and get hot! Yay, soup! Once you take your soup out of the microwave and set it on the counter while you get some crackers, the soup will start releasing heat back into the atmosphere. You see this as steam. The soup and the air temperature are trying to equalize.

The soup example gives us a good look at one of the fundamental differences between classical and quantum physics. Microwave ovens make use of one of our favorite electromagnetic waves to artificially excite the atoms in our food. It is not the application of heat in the traditional sense like a fire or a stove burner. In a way, every time you turn on your microwave, you're using quantum physics in your kitchen. When you take the soup out of the microwave, you see classical physics in action. The law of thermodynamics defines thermal equilibrium. Temperatures will always try to sort themselves out. This is achieved through convection, conduction, or radiation. Convection would be heating your soup by putting it in a pan on a gas stove. Conduction would be putting the same pan on an electric stovetop. Radiation is putting the soup in the microwave. Make sense?

These laws also all help us to understand the states of matter, which applies to both classical and quantum physics. As we know from primary school, there are three main states of matter- solid, liquid, and gas- and the fourth state, plasma. Most substances on Earth and throughout the known universe come in one natural state and can only take the other states (or phases) when acted upon by an outside energy, such as heat, cold, pressure, motion, etc. Sometimes we are humanly capable of causing these changes, and sometimes they occur naturally. Let's go back to our soup for a second.

Because we are so fancy, our can of soup is one of those concentrated tins that you have to add water to after you open it. So, when you first crank up the can opener and plop it into your saucepan (the microwave is broken for this example), it's solid. We all know that noise, don't kid yourself. Even if we didn't add the two recommended cans of water to start thinning it out, the soup would begin to "melt" and liquify after you turn the heat up under it. Applying heat changes the state of the soup from solid to liquid. Now, imagine that your stove is capable of heating that soup until it completely evaporates. Now your soup is gas. Yum! Now it would be really impressive if you could turn your cheap lunch into plasma-that's the state of matter where a gas is stripped of its electrons and seemingly takes a near-solid state. An example might be a neon bar sign, where the neon is ionized inside sealed vacuum tubes to create light. Or, you know, our Sun. That's made of plasma, too.

Obviously, we're being a little ridiculous for the sake of the example. Still, it is essential to see the states of matter at play in everyday life so we can understand how scientists use these basic laws to understand the parts of the universe we can't see, like the tiniest of subatomic particles and the vast celestial bodies in the outer reaches of space. Classical physics is useful but has its limitations because of its solidly "predict and prove" nature. Quantum physics goes beyond the "how does this work?" to explore "why does this work?" and puts an awful lot of stock into the unknown. Because of this, even quantum physics has to have its feet planted firmly in the things we DO know so that we can figure out the things we don't. Along with these physical laws, there are some numbers that help quantum physicists stay grounded as they perform their work.

Measuring Large Numbers of Tiny Things

Do you like Avogadro? Isn't it great for making guacamole? Oh, sorry, that's avocado, the green stone fruit used in traditional North and Central American cuisines. Avogadro was a 19th-century Italian chemist who was also a scholar in mathematics, physics, and electricity. He had two separate tenures as the head of the Academy of Sciences in Turin during his career and focused much of his work on the behavior of gases under a variety of chemical and physical conditions. He was especially fascinated by being able to predict that behavior based on the quantity and weight of the substances that he was working with, and the results of his work are still used by classical and quantum physicists and chemists today. What was that lasting contribution to science? The cleverly named Avogadro's number, or Avogadro's constant.

Avogadro was a pioneer in a time when there was yet to be a solid model of the atom or understanding of atomic and molecular weight and behavior. He studied the actions of gases, was able to come up with startling accurate chemical formulas, and spent many research hours determining a way to measure the number of atoms or molecules in a finite quantity of gas. He theorized that under the same conditions (a temperature of 0 degrees Celsius and at normal gravitational pressure), one liter of all gases would contain the same number of particles. This number is what we now know as Avogadro's number:

$$6.022 \times 10^{23} = \text{one molecular gram (or mole)}$$

This number aids scientists in calculations in both quantum and classical physics, chemistry, and mechanics. Unfortunately, he didn't live to see his theorem become law. It was proven in 1858, two years after his death, by his friend and colleague Stanislao Cannizzaro. Cannizzaro was able to show that Avogadro's hypotheses about molecular weights, volumes, and amounts were solid, cementing Avogadro's legacy in the scientific world. If you ever have a difficult time remembering Avogadro's constant, here's a little hint for you:

A mole is a number, or have you heard? It contains six times ten to the twenty-third.

It's even better if you sing a little tune to help make it stick.

Planck and His Constant

Moving forward into the next century, theoretical physicist Max Planck gave the world another vital constant- named after him, of course. It's often said that the difference between classical and quantum physics is the need to use Planck's constant in your calculations. Planck was among those working on black body radiation around the turn of the 20th-century, and the scientists were in a bit of a conundrum. They had figured out how black body radiation worked, but they were still struggling with calculating how much energy a body was absorbing to compare how much it was emitting back in the form of other waves. They needed a constant to complete their mathematics. Planck to the rescue!

The biggest problem with the researchers' previous calculations was that the energy always seemed to behave differently after a certain point in time and at frequencies higher than 10^5 GHz (gigahertz, a measure of wave frequency). This inconsistency became such an issue that it was dubbed the "ultraviolet catastrophe," which suggests that these practical men of science were nearing the end of their rope trying to figure it out. That is a tremendously dramatic name for a scientific shortfall. Anyway, the concern was that the energy emissions from a black body under controlled circumstances followed along with everything that classical physics said that it should until the energy hit that magical threshold of the ultraviolet catastrophe. As soon as the energy emissions reached the ultraviolet part of the electromagnetic spectrum, all hell broke loose. The researchers were clearly at their wits' ends.

Planck began to look at the problem from a different angle. What if the energy was acting exactly as it was supposed to, based on wave-particle duality? And if that were the case, how could they begin to predict its behavior with mathematics? Planck was confident he could devise a way to do this and started playing around with the numbers. If he used classical theories, then the energy emitted by a black body should rise in a steady curve over time, but that's not what they were seeing. Planck's constant factors in one thing that classical physics did not- and that's proportionality. By comparing the proportion of energy over the frequency of the waves being emitted, Planck not only revolutionized physics, but he solved the ultraviolet catastrophe! What a hero!

Planck's constant is symbolized in physics equations as *h*, and it represents this number:

$$6.62607015 \times 10^{-34} \text{ J·s}$$

Let's break down what this means, especially the unit of measurement, J·s because it can be a bit confusing. The "J" stands for the SI unit Joule, which measures one unit of energy. The "s" is for seconds elapsed. The confusion with Planck's constant lies in the measurement not being Joules per second, but in fact, a new measurement devised by Planck of the Joule-second. He used the combination of a unit of energy and a unit of time to create a single unit that would denote action or momentum. That's why Planck's constant is of such utility to quantum physicists. It gives them a unit of measure that aligns with the quantum theory that all matter is constantly in motion.

For context, Planck's constant isn't just some number that floats around alone and does nothing. It is used in a variety of calculations, including the one that started it all-the one that predicts the behavior and momentum of black body radiation. It's also used to determine spectral radiance, which calculates the amount of radiation a black body will emit. Bohr also used Planck's constant when working on his updated atomic model in the 1910s. It also plays heavily into the calculations of the photoelectric effect and other quantum physics principles, so let's take a look at those now.

Origin of the Theory of the Photoelectric Effect

If there were to be one defining moment between classical and quantum physics, it would have to be a toss-up between the development of Planck's constant and the theory of photoelectric effect. Let's take a moment to define "quantum," which will shed a lot of light on the photoelectric effect (Hah! Shed light. On the photoelectric effect. Ahem. Sorry.) A quantum – plural "quanta"- is a singular, discrete packet of measurable energy. It describes the minimum amount of a substance involved in a physical interaction. These interactions are classified as gravitational, electromagnetic, strong, and weak and are based on the principles and laws of both classical and quantum physics.

These fundamental interactions are centered on the unbreakable laws of physics. Gravity is the natural force that draws all objects with mass or energy towards each other. We base the measurement of gravity on what we see on Earth (1G). Electromagnetic interactions are those between charged particles, like electricity, and electric or magnetic fields. Strong and weak interactions are those within a substance itself; strong interactions are what hold an atom together, and weak interactions are those that can let an atom fall apart, like radioactive decay. These fundamental interactions form the basis of all physical behavior.

Getting back to defining what makes a quantum a quantum, it is the minimal amount needed to perform one of these interactions. Let's say you are having a water cooler conversation with a particularly dull colleague, and they are droning on about something they did over the weekend. In order to participate in the chat, you dredge up a single utterance- "Wow, that's great!"- and extricate yourself to go back to your desk. You've offered a quantum of small talk, i.e., the minimum quantity of words you need to be part of an interaction. That's quantum physics in a nutshell. It aims to determine the smallest possible amount of material necessary to perform the universe's fundamental interactions and explain how and why the material behaves the way it does.

So what does this have to do with the development of the theory of the photoelectric effect? When Planck was working on coming up with his constant, he was trying to explain the behavior of radiation from a black body. Other scientists were trying to figure out how light on the visible spectrum played into all this and measure the number of electrons that were released from other surfaces (non-black bodies) when light was shone on them. The two experiments came crashing together when Albert Einstein began to realize that light exhibited wave-particle duality in the first decade of the 1900s. The resulting theory would become the law of the photoelectric effect.

Prior to this, light had always been treated solely as a wave. Electromagnetic radiation was measured in frequency and wavelength, never in mass or quantity. Scientists like Henri Becquerel's father, Edmond Becquerel, and Heinrich Hertz were responsible for standardizing the measurement of electromagnetic waves in this manner due to their mid-1800s work with the EM spectrum and photovoltaic effects. But science evolves rapidly, and by the turn of the century, researchers were looking for new answers to their questions about the behavior of visible light and other electromagnetic radiation.

Einstein believed that light could be measured as a particle, which theoretically would make it a substance and not a wave. (Ah, good old wave-particle duality, sigh…). By measuring light this way, it could be "quantized" or assigned a value that would help researchers calculate the number of electrons released by a surface when light was applied to it. In other words, they could now predict the photoelectric effect. Einstein called his particles of light "photons," a term which we still use today. Planck's constant and Einstein's research came together to form the Planck-Einstein Relation, shown as the equation:

$$E = h\nu$$

where E is the kinetic (moving) energy of the photoelectrons, or photons, this energy is equal to the frequency of the light wave multiplied by Planck's constant. This equation can be manipulated to explain any of the missing variables, and it was a true "lightbulb" moment for the early quantum physicists. If light itself exhibited wave-particle duality, then anything was possible. They just needed to keep digging to begin to explain it all.

Waves, Particles, and Sub-atomic Stuff

Keep digging; they did. So many things were happening at once in the young discipline of quantum physics in the early 20th century. While Planck was developing his constant and fixing the ultraviolet catastrophe, Einstein was working out the photoelectric effect. At the same time, Bohr and Rutherford were figuring out how to make more accurate atomic models, and other scientists were also putting together some really exciting stuff. One missing piece was filled in by quantum chemist and mathematician James Chadwick in 1932, and that was his proof of the existence of the neutron.

Something that had been puzzling physicists and chemists are they strove to discover, weigh, and classify new elements and finalize the atomic was the inability to explain why atoms held together the way they did. Even as they began to understand atomic radiation and radioactivity more, there was still something out of their grasp. Chadwick believed that there had to be another particle within an atom to balance out the charge between the electrons and the protons and give each element its proper mass. He set out to uncover exactly what that missing bit was.

Chadwick was more than a quiet scientist, to be fair. He had traveled to Germany as a young man to study radioactivity under Hans Geiger (for whom the famous radiation counter is named) and ended up embroiled in World War I. The Englishman was promptly tossed in a prison camp for, well, being English. He managed to rally his incarcerated countrymen into forming a makeshift science club, wrangled some toothpaste containing radioactive elements into the camp, and found a way to conduct experiments with a crude electroscope made from wood and leftover tin foil. This guy was serious about his scientific pursuits. Fast forward a few years,

Chadwick is back in England, working in academia, when he decides to tackle atomic theory. He was up to the task, for sure.

To prove that there was more going on in the nucleus of the atom beyond just protons, Chadwick set up a series of experiments using beryllium, notably stable, and polonium, notable radioactive. He wanted to explain the difference between atomic number and atomic mass, and other colleagues were working with these elements and not getting the results they wanted. To others, the instability and radiation from the beryllium after being bombarded with polonium must signify that the polonium was causing the beryllium to become unstable. To Chadwick, it seemed that something else was afoot, and so he tried their experiments for himself.

After playing around with the set-up for a scant two weeks, Chadwick tentatively decided that the polonium itself couldn't be responsible for the behavior of the beryllium, which was releasing too many particles for them all to be protons or electrons. What if there was another sub-atomic particle at the heart of the atom? Wouldn't this also explain the discrepancy between atomic weight and atomic number? He theorized the existence of the neutron early in 1932, and by 1934, the theory had been proven correct. With his discovery, Chadwick made it easier for scientists to classify new elements, create an accurate atomic model, and paved the way for a slew of other atomic experiments. This included the bombardment of uranium, which researchers soon found released a profound amount of energy very quickly. This, for better or worse, led to the development of atomic weaponry, and Chadwick himself even collaborated on the Manhattan Project during World War II. He was also the Nobel Laureate in 1935 for his findings.

On the subject of wave-particle duality, we would be entirely remiss if we didn't mention the guy who began to make sense of it all, French physicist Louis de Broglie. His 1924 thesis on the wave nature of electrons contributed to the basis of wave-particle theory and incited the growing snowball that would become the avalanche of quantum physics. Louis de Broglie, the 7th Duc de Broglie, didn't let his aristocratic status stop him from becoming a renowned researcher, and he got much of his start in physics from his military service in World War I. He and his colleagues were in charge of monitoring the radio communications based at the Eiffel Tower in Paris and later developed and installed radio systems for use on submarines. Although de Broglie loved his communications work and found the radio transmissions fascinating, his true scientific passion lay in academia and particle physics, which he returned to after the war.

In the 1920s, de Broglie spent time in the laboratory of his brother Maurice, also a physicist, working with x-rays and studying the photoelectric effect. He was enthralled with the inner workings of these waves but longed to be able to explain their behavior as he observed it. After experimenting with x-rays, de Broglie concluded that there had to be a way to link waves and particles. Latching onto Einstein's theory of the photoelectric effect, de Broglie zeroed in on the notion that each quantum of light must have something telling it how to act, which he called the "pilot wave." These pilot waves would dictate what direction the particles would move and at what frequency and wavelength. The purpose of his research into pilot waves was to prove that he could predict the wavelength of any moving object, including subatomic particles like electrons and photons. He subsequently came up with the de Broglie wave equation to prove his theory mathematically. The equation is as follows:

$$|$$

Here, you can see Planck's constant at work. The Greek symbol lambda represents the wavelength we are trying to determine. This is done by dividing Planck's constant by the momentum of the particle, which is calculated by multiplying the mass times the velocity. Therefore, if you know how much a particle weighs and how fast it is going, you can use Planck's constant to discover the wavelength of that particle. Bam! Wave-particle duality spelled out in one simple equation. Physics was forever changed by the curiosity of a French duke who couldn't stand thinking that waves and particles were two different things. He knew there was a relationship, and he set out to prove it, winning himself the 1929 Nobel Prize for doing it, too. The de Broglie equation is still used today in all levels of quantum mechanics and physics because of its efficacy in determining wavelengths.

So Much Uncertainty

The last big concept we'll look at before moving on to a deep dive into a couple of physics' most renowned modern scientists is that of uncertainty. Maybe you've heard of the Heisenberg Uncertainty Principle and just never been able to wrap your brain around it. That's more than fair because it's a bit mind-boggling. In its simplest terms, the uncertainty principle states that the more precisely you know a particle's location, the less certain you are about its velocity and vice versa. Um, what? Shouldn't having exact coordinates and measurements mean that we're *more* certain about something?

In plain language, the uncertainty principle is meant to point out how difficult it is to track and predict the movement of individual particles. By the time you've calculated the position of a particle, it has already moved on, and even if you know how fast it is going, it will be hard to predict where it will be at any given time. Heisenberg also indicated that this uncertainty grows with time and distance from the origin point of the particle-wave. He wanted to find a way to make sure that physicists knew what they were up against when they attempted to take measurements of quantum particle waves. Heisenberg used a reduced Planck's constant (the two-dimensional rather than the three-dimensional version) to indicate that the change in position and the change in momentum is greater than or equal to the reduced Planck's constant divided in half. The simplest mathematical form of the uncertainty principle looks like this:

$$\Delta\chi\Delta\rho \geq \frac{\hbar}{2}$$

The uncertainty principle calls into play one of the most essential things about quantum physics versus classical physics- that even though we know how something is behaving, it doesn't mean we can always predict what is going to happen next. The objects around us are moving in waves right now. The atoms are alive with a motion so imperceptible to the naked eye that you would never know, but it's true. Heisenberg wanted to be able to account for this unpredictable constant movement with math.

The other thing to keep in mind about the uncertainty principle is that it is often confused with what's called the "observer effect." This is something that all scientists are concerned about when performing any experiments; they don't want their very involvement to be the reason the results are skewed. The uncertainty principle is going to manifest whether someone is observing it or not- sort of like how the tree still makes a noise if no one is around to hear it. The observer effect could be something as simple as turning on a light in a laboratory or getting fingerprints on a sensitive object and contaminating the sample. Scientists must always take precautions to guard against the observer effect when they are working, but there is nothing to be done about the uncertainty principle. It's going to happen no matter what.

One of the scientists who we haven't mentioned much in these first couple chapters but who warrants looking at based on his thoughts on many of our topics, including the observer effect, is Erwin Schrödinger. In the next chapter, we're going to talk about this eccentric scientist's impacts on early quantum physics, his famous thought experiments, and his legacy. Let's take a look!

Chapter 3: Schrödinger, His Theories, and His Cats

One of the most famous names in quantum physics is Erwin Schrödinger, although it's likely you've heard of him more for his thought experiments than his actual science. Schrödinger was an Austrian-born scientist who emigrated to Ireland and finished his career there, after a little run-in with the Nazis leading up to World War II. He's also famous for his "feud" with Einstein, due to Schrödinger's so-called failing to embrace the fundamentals of quantum physics as early as some of his colleagues. We'll talk about his infamous cats and his philosophical pursuits in a little while, but first, let's see what he did in the laboratory over his storied career.

Schrödinger's Quantum Theory and Unified Field Theory

In the 1910s, Schrödinger became acquainted with the works of Planck, Einstein, and others, but wasn't terribly interested in giving up classical physics just yet. That all changed when he developed tuberculosis and spent some time in and out of sanitoriums regaining his health. While convalescing, the scientist started to warm up to the ideas placed before him, especially quantum theory. He was fascinated with the possibility that the electromagnetic spectrum could have differing effects on a wide variety of elements and wanted to discover if he could find a universal way to predict the behavior of electrons.

Schrödinger played around with these thoughts and came to the conclusion that the only way to do this would be to be able to predict the nature of radiation itself. While he couldn't find a way to do this, the articles he published proposing these concepts opened the door for a new brand of theoretical physics and did lead to a few concrete advancements, such as wave mechanics and an interesting new atomic model, which we'll get into shortly. Schrödinger's other contribution to theoretical physics was his attempt to create a unified field theory, which is sort of the Holy Grail of quantum physics. A unified field theory, or UFT, had been attempted by Einstein when he was working on his theory of relativity and is still a matter of great debate among quantum scientists.

A unified field theory would bring together all the fundamental interactions and relationships in quantum physics under one set of proven laws. This would mean that researchers would be able to predict the behavior of all matter based on mathematical proofs and observable actions with no deviations. Matter would behave according to these laws, and we'd be able to connect the dots between interactions, electromagnetic fields and waves, particles, and spacetime. There have been more than a few attempts to write and prove a unified field theory for quantum physics (but classical physics has one…), but no one is yet to succeed. One such popular attempt is called "the theory of everything," and let's face it, that does sound a bit pretentious.

Congratulations! It's Wave Mechanics

One thing that Schrödinger was successful in was becoming the father of wave mechanics. His theories on the behaviors of particle waves gave birth to this new sub-discipline in the mid-1920s. The basis of his theories was rooted in the premise of the behavior of a hydrogen atom in a system independent of time constraints; Schrödinger wanted to know what would happen if time was taken out of the equation when predicting the wave-particle behavior of the atom. In the resulting rapid-fire series of four papers, the physicist laid out his predictions and gave the world the first look at his now-famous equation; the second paper gave the equation an edit to account for harmonics within the system.

The third and fourth papers in the series were devoted to showing the world how to compare his equation to work involved in the uncertainty principle, and taught his colleagues how to plug in the proper complex numbers directly into his equation to avoid having to calculate so many derivatives. This series of papers is still considered one of the greatest scientific accomplishments in modern science. It paved the way for physicists to begin studying wave behavior in a much more controlled and accurate manner.

It also marked the beginning of some seriously complicated math, and many believe Schrödinger's equation to be the cut-off between classical and quantum physics. Do you see now why it's so difficult to pinpoint one defining moment? Even Schrödinger himself wasn't thrilled with what he'd done, once it was over and published. He recognized that he had caused a giant rift between classical physics and the new quantum discipline. As someone who loved the principles of classical physics, he never wanted to be as right as he was- but once his research was published, there was no going back.

Schrödinger's Equation and Atomic Model

What exactly was so special about Schrödinger's equation, other than the fancy numbers? And why did it have such a profound effect on quantum physics? To begin with, it is a partial linear differential equation, which means it has a lot of moving parts in order to get to its solution. Heck, that's a lot of words just to *describe* an equation. In classical physics, there is Newton's second law of motion, which you may recall is shown mathematically as F=ma or force equals mass times acceleration. It predicts the movement of an object as it speeds up, proportional to its mass. Think of Schrödinger's equation as the quantum physics counterpart to Newton's second law. Yes, it was that groundbreaking and that important. Below here, you will see the equation:

$$H(t)|\psi(t)\rangle = i\hbar\frac{\partial}{\partial t}|\psi(t)\rangle$$

Yeah, we know. This is a lot to unpack, so we're just going to pay the baggage fees on this one and sent it on its way. After all, this is a book for beginners. This sucker took F=ma and turned it on its classical physics' head. But for those who are capable of looking at Schrödinger's equation and doing the math, it is crucial. Without this equation, scientists might still struggle to predict the behavior of wave functions over time. With the equation, a little simple math gives them all the answers they need. It's said that with this one calculation, Schrödinger flung open the door of quantum mechanics, and it was never able to be shut again.

Schrödinger was also intrigued by the thought of updating the atomic model to account for the wave behavior of electrons as they orbited the nucleus. After careful analysis, the scientist was able to create the first truly three-dimensional, accurate model of the atom, using his wave mechanics theory and equation. Schrödinger's model of the atom includes an electron cloud, moving in and out around the nucleus in a wave pattern. Using this model, scientists can predict where the electrons might be at any given time. This differed from the earlier Bohr model that showed the electrons in set, layered orbits that did not fluctuate due to wave-particle duality.

Yes, Yes, the Cats...

Early in his career, Schrödinger was interested in sticking to classical physics, and as he aged, he made the leap into quantum physics and theoretical physics. Not only was he an outlier among his colleagues for waiting so long to join the game and being sad for himself that he did (despite the accolades and the Nobel Prize), but he was also a certified odd bird. He had a very untraditional marital arrangement, marked publicly by emigrating to Ireland with both his wife and mistress along with the children that he fathered with both women. He wrote a premise on genetics, which inspired a whole generation of geneticists and led to the discovery of the human genome. Schrödinger also authored a treatise on the nature of colors on the visible light spectrum and color perception in humans. He had a lifelong curiosity for the intersection of science and philosophy.

His most famous thought experiment is that which is known as Schrödinger's cat. In this exercise, Schrödinger theorized that a cat, locked inside a box with a deadly poison, can be supposed to be both alive and dead, at the same time. Yes, it's a paradox, and it was designed to be. One cannot tell what state the cat is in without opening the box and looking at it. But what if opening the box changes the predicted outcome? Schrödinger first introduced the paradox in a series of debates he held with Einstein. The thought experiment was his quiet way of railing back against certain interpretations of his quantum mechanics theories.

Schrödinger wasn't happy with the so-called Copenhagen Interpretation of his wave mechanics. This school of thought, led by Niels Bohr of atomic model fame, stated that wave function would collapse because there was no way to observe the nature of waves without interfering with them and causing disastrous results. Bohr and his colleagues were convinced that every act of measurement would affect the experiment until the outcomes were invalid. Schrödinger argued that no collapse had ever been physically observed, so how could Bohr prove his interpretation? Out of this frustration, the Schrödinger's cat thought experiment was born. He wanted to show his fellow scientists that they were being ridiculous to think that, yes, while observer effect is a thing, the mere act of looking at something would cause a drastic impact on its status. To do so, he put an imaginary cat in a box and began asking people if they thought it was alive or dead.

While this does seem a little ridiculous, it had a profound effect on the scientific community and the future of quantum physics. From Schrödinger's cat came the theory of quantum entanglement. Schrödinger himself was forced to concede that maybe his quantum wave mechanics couldn't explain everything all the time, and he began to recognize that maybe what his Copenhagen-interpreting friends were seeing wasn't the breakdown of wave function, but the tangling up of waves as they crossed each other. Like children playing cat's cradle and getting their yarn hopelessly knotted up, maybe electrons were getting their waves tied up together, making their paths indistinguishable. The theory of quantum entanglement brought a new aspect to wave mechanics and made it possible for scientists to know what to look for when studying the behavior of subatomic particles.

Erwin Schrödinger died of tuberculosis in 1961 and was laid to rest in his native Austria. His wave equation is engraved above his tomb, and his wife was buried with him after her death in 1965. No word on what happened to the mistress. For what it's worth, Schrödinger always marched to his own drum. He joined the fray of quantum physics if and when it suited him, lending his brilliant insight and then bowing out to pursue his other passions whenever he pleased. He didn't give a hoot about the Nazis invading his homeland, and he told them so, he flaunted his open marriage in public in the 1930s and 40s, and he gave the world a scientific way to talk about dead cats. In one of his last public appearances in the 1950s, he even declined to talk about nuclear power, as he'd been asked, and spoke on philosophy at length until his audience protested. Schrödinger was fabulously gifted, delightfully weird, and tremendously impactful on the world of quantum physics. Almost as much as the next guy… Let's take a gander at Einstein and his contributions.

Chapter 4: It's All Relative - The Genius of Einstein

When many people think about Albert Einstein, it's almost always as he was in his elder years, with crazy hair, touring the world giving lectures and engaging in lively debate. While that is an accurate picture of the renowned scientist, the truth is that most of Einstein's seminal work was completed and recognized when he was a much younger man. And even though he was absolutely brilliant, he wasn't always on course with his ideas. For the next couple of chapters, we'll take a look at the groundbreaking concepts, the wins, the losses, and the feuds that made Einstein both one of history's most revered and most controversial figures.

A Portrait of the Scientist as a Young Man

Albert Einstein was born in 1879 in the former German Empire, and his German heritage would play heavily into decisions that would shape his fate throughout his life. The son of a salesman turned engineer, young Einstein and his family moved to Munich where his father and uncle operated an electrical supply factory. From an early age, Einstein was fascinated by mathematics and scientific processes. Although the family was Jewish, Albert received his elementary education at a Catholic school and then attended a secular secondary and high school.

When his family business failed to secure a significant municipal contract in Germany, the elder Einsteins chose to relocate to Italy for new opportunities. Albert, desperate to begin higher-level studies, decided not to accompany his family to Milan, but a falling-out with authorities at his school soon sent Albert packing to meet his family in Italy. Apparently, the 15-year-old future Nobel Prize winner was unsatisfied with the institution's teaching methods, but his teachers weren't exactly thrilled with him vocalizing his concerns.

Einstein longed to be admitted to a university but continued to be thwarted by age and immaturity. He wrote lengthy essays on magnetic fields and electrical energy and taught himself Euclidean geometry, algebra, and Pythagorean theorem. He was also fascinated by philosophy and music, and enjoyed playing the violin and reading Immanuel Kant. Einstein tried to gain admittance into the Swiss Federal Polytechnic School at 16, and while he excelled in math and science, he was rejected for low scores on the general knowledge sections of the entrance exam. It seemed that the scientific prodigy would be held to the same standards as his peers, despite his proclivity for the more advanced subject matter.

Einstein did well enough on the exams the next year to be admitted to the institute, where he focused his studies heavily on mathematics and classical physics, earning a teaching diploma and meeting his first wife, Serbian scientist Mileva Marić, who was one of the only young women there studying in the same field. It was also during his time at school that Einstein renounced his German homeland and became a Swiss citizen. He would have no regrets about this decision, as the young scientist was a pacifist, and Germany was prone to war. Although the Swiss Army also had conscription, he managed to get a health exception. Therefore, Einstein would never be forced to take up arms under the flag of any nation that he called home.

Even after earning his teaching certificate, Einstein found it difficult to attain a position, and so he took a job as a review clerk in the Swiss Patent Office. This is where the brilliance of Einstein began to shine. He was a scientist with no laboratory, working at a dead-end government job (where he was passed over for promotion because of his deficiencies in working with new technology!), recently married, and beginning a young family. Yet, somehow, in his SPARE TIME, Einstein started developing and writing what would become some of the most significant breakthroughs in the fairly new field of quantum physics, along with completing his thesis to earn his Ph.D. from the University of Zurich. And some people can't walk and chew gum at the same time!

Miracle Year: Unlocking the Secrets of the Photoelectric Effect

Einstein worked at the patent office from 1902 through 1909, and in the midst of that, he had what is known as his Annus Mirabilis, or 'miracle year' in 1905. In those twelve months, he wrote his doctoral thesis on new methods of determining molecular dimensions, as well as a series of four mind-blowing papers that turned classical physics upside down.

Included in the papers produced during the miracle year was a study in Brownian motion, which describes how to measure and predict the seemingly random movement of particles suspended in a liquid or gas. Einstein's calculations and explanations gave chemists and physicists a new way to think about the substances they were working with. Using Einstein's recommendations, researchers were able to conduct their experiments with greater accuracy to achieve more substantial, pinpointed results.

One of the other mysteries unraveled in Einstein's 1905 papers was the long-standing discussion on the photoelectric effect, which we talked about briefly earlier, so let's take a closer look now. If you'll recall, the photoelectric effect is what happens when a stream of light or other electromagnetic wave is directed towards a solid object, and particles begin to scatter from the surface. The biggest problem with discovering exactly how the photoelectric effect works is that the behavior of the light and the behavior of the electrons being released from the solid surface didn't match up with what should happen, according to classical physics and mechanics.

Einstein looked at the work being done by classical physicists and early quantum physicists, especially Max Planck, and realized that Planck must be on to something with his quantum theory. Planck believed that all electromagnetic waves, including visible and invisible light, were made up of small, individual packets of energy, which he and his colleagues dubbed quanta. We compared quanta earlier to the minimum interaction you need at the work water cooler, and it's true. A quantum is a tiny little packet of energy, bundled up and doing its thing with its quanta friends.

As Einstein studied the photoelectric effect, he began to feel even more strongly that Planck was correct about quanta, but none of Planck's nor Einstein's research lined up with classical physicists' idea of how the effect worked. Most classical physicists believed that it was changing the intensity of an electromagnetic wave that determined how many electrons were freed by contact, but it was Einstein working off of Planck's research that cemented the theory that it is the frequency of the waves, not the intensity or the duration of the waves. In other words, a high-frequency wave such as an x-ray will cause more electrons to be freed than a low-frequency wave like a radio wave, regardless of the focus or length of time the waves are directed at a surface.

Classical physics tells us that the stronger something is and/or the faster it is moving, the more force it carries, but that's not the case with the photoelectric effect, and hence, there's the reason for the rift it caused between classical and quantum physics. Force, speed, and intensity have nothing to do with the photoelectric effect, and Einstein, in conjunction with Planck's ideas, was able to prove that. The conclusion that wavelength was responsible for the magnitude of the photoelectric effect also proved wave-particle duality in a practical setting because Einstein was able to use Planck's constant to calculate the energy of the electromagnetic waves as they hit the solid surface. As a reminder from earlier, this equation is:

$$E = h\nu$$

Or the energy equals Planck's constant times the frequency of the wave. Experiments over the years have played with all types of electromagnetic waves and solid surfaces, from ultraviolet, visible, and high- and low-frequency waves from either end of the spectrum being streamed onto metal and non-metal surfaces, as well as synthetic compounds. The results have always shown that Einstein and Planck were right about everything: Electromagnetic waves have wave-particle duality, these waves are made up of individual packets of energy known as quanta, and it is the frequency of these waves, not the force of them, that causes the photoelectric effect. Einstein would win a Nobel Prize for his work on the subject.

Special Relativity, Mass Equivalency, and General Relativity

Now that we've covered two of the four miracle year papers, you're probably wondering when the heck we're going to get to the good stuff like relativity. The wait is over, so buckle up and let's go for a ride through space and time and all that jazz and discover what was so special about Einstein's other 1905 research articles and how they led to even greater things. We're talking, of course, about special relativity and mass equivalency- and from these last two *annus mirabilis* papers also sprung the theory of general relativity.

Einstein developed the Theory of Special Relativity as a way to explain the unexplainable, especially why classical physics, namely Newtonian physics, couldn't account for the movement of things going near or at the speed of light. Classical physics had no answer for why particles that move so quickly seem almost to be standing still. Einstein, because he always wanted to have a solution for everything, came up with one. And because it's Einstein, the concept, of course, came to him in a dream. Yes, the man literally was able to conceptualize advanced theoretical physics in his sleep, or more specifically, in a "thought experiment" that he first had when he was just 16 years old.

This particular thought experiment focused on the wave-particle duality of light and how we perceive its movement through space; more specifically, as the story goes, Einstein dreamt about riding a beam of light through the vacuum of space. (It should be noted that Einstein's detractors scoffed at the idea of a teenager dreaming up such advanced theories, but it does make for a fascinating tale!) What the young scientist was trying to figure out was why light particles didn't behave according to the laws of classical physics both in and out of a vacuum. As Einstein was riding his light beam, he was also observing a parallel beam of light alongside him, meaning the relative speeds should have been zero, but they weren't! Why?

Newtonian physics tells us that motion must follow these laws:

> **1-** An object at rest stays at rest, and an object in motion stays in motion until acted upon by an outside force.
>
> **2-** Force is equal to mass times acceleration for objects of unchanging mass; otherwise, force is equal to change in momentum over the change in time.
>
> **3-** Every action has an equal and opposite reaction.

But electromagnetic waves, especially light, which can travel unimpeded through a vacuum, don't follow these rules, so what gives? Einstein was determined to figure it out, and by 1905, he was ready to give his answers to the world. The key to special relativity lies in not thinking of time as a construct but as its own measurable dimension- an additional axis, if you will. Since light is the fastest thing in the known universe, and nothing can travel faster than the speed of light, that would seem to explain why anything approaching the speed of light would seem to slow down and gain mass, making it impossible for that object ever to exceed the speed of light.

There's also the problem of linear observation. What a beam of light looks like to a person standing at one fixed point is vastly different from what it looks like to someone standing at a different point, or, as in Einstein's thought experiment, riding along on a parallel beam of light, where the speeds of the two parallel beams should be the same- or relative. Think about standing on a train platform in a fixed position and watching a train speed by you. To you, an observer who is not moving, the train seems to be going pretty darn fast. But if you were also moving as a passenger on a train traveling on a parallel track, the first train's speed would be harder to gauge because it would become relative to yours. The first train hasn't adjusted its speed at all, but it will appear to move faster or slower based on the observer's own moving or stationary viewpoint. A person on the platform and a person on the parallel train will see different things, but the first train is still moving at the rate it is moving, regardless of who is watching it from where. The train's speed is measurable, but it is also relative to the observers around it.

Let's break this down into smaller pieces to make it easier to digest. If special relativity seems counterintuitive to you, that's because it is. It applies to a very specific action of electromagnetic waves and the photon particles which make them up and laughs in the face of classical physics. Special relativity also accounts for the phenomenon known as time dilation because it treats time as its own dimension. Einstein realized that even as objects approach the speed of light, unable to surpass that speed due to their own increasing mass holding them back, these objects also seem to slow down time itself.

This is why the modern case of identical twin American astronauts Mark and Scott Kelly is so fascinating and so telling. Both men were given comprehensive health exams before and after Scott spent a year on the International Space Station, which moves at an average speed of 17,130 miles per hour. When Scott returned, and the men were examined again, it was determined that Mark had aged more over the course of the year on Earth (which rotates at approximately 100o mph) than his brother, who was moving much more quickly through space.

Einstein's special relativity theory also gave the world one of its most famous equations. We're talking, course, about:

$$E = mc^2$$

But what does this simple mathematical statement tell us about how the universe works? This is the equation that shows mass equivalency, and its simplicity is what makes it so brilliant. Einstein was able to look at a huge problem (light not behaving as classical physics says it should), determine the elements that make up the problem (light being able to move through a vacuum and nothing being able to travel faster than the speed of light, which is 186,000 miles per second), and figure out how to calculate the energy assigned to objects even as they approach the speed of light, gain mass, slow down, and dilate time. One equation to do all that!

In the mass equivalency equation, we see the E representing energy is equal to the mass (m) times the speed of light (c) squared. Mass equivalency gives us the means by which to express the energy of a particle-based on its mass and speed, which would, much to Einstein's chagrin, become integral in the development of atomic weapons, but also in the rise of other nuclear power applications. Mass equivalency also works in the applications of other systems that require the calculation of mass and potential/kinetic energies, such as chemical reactions and controlled explosions. The mass equivalency equation was the subject of the last of the annus mirabilis papers and was published in November of 1905.

We're going to dive into general relativity in a moment, so let's recap the big points from special relativity and mass equivalency to get our heads on straight before we move on. The special theory of relativity is essentially a way to explain the movement of photons and other electromagnetic quanta as they travel in a straight line through a vacuum. It puts a speed limit on how fast these particles, and any other for that matter, can travel. Simply put, nothing can move faster than the speed of light. The theory also addresses how things that are in motion appear to be moving at different speeds based on the viewpoint of the observer and why adding time as its own dimension in these observations results in the phenomenon of time dilation. Mass equivalency tells us how to translate the mass of an object into the energy it contains using the speed of light as a constant. This equivalency opened the doors to a whole new perspective on quantum mechanics and quantum chemistry.

After his onslaught of advancements in 1905, would you believe that Einstein STILL couldn't find a teaching position? True story! He continued to work at the patent office until early in 1909, but that didn't mean that his wheels weren't turning on the next big thing. During the ten years between releasing the theory of special relativity and the theory of general relativity, Einstein did manage to get out of the patent business, teach in Bern, move back to Germany to take a position lecturing in Berlin, have a third child, become separated from his wife, teach in Prague and become an Austrian citizen, reconnect with two old flames (more on Einstein's unconventional love life in the next chapter), and move *back* to Zurich to lecture at his alma mater. Phew! Anyway, by 1915, Einstein had crammed in a whole lot of upheaval, a whole lot of lecturing, and a whole lot of thinking about relativity, again.

The problem that Einstein was working through was that special relativity is, well, special. It's very specific to the action of electromagnetic waves moving in a straight line through a vacuum.

That's a pretty detailed set of circumstances. Einstein wanted to figure out how relativity applied to everything else in the universe that didn't seem to fall in line with classic, Newtonian physics. Why do the planets revolve around the Sun? How do black holes absorb all the energy and mass around them? How does time work as the fourth dimension? Einstein wanted to find all the answers and bundle them neatly in one general relativity package.

One problem that presented itself is that classical physics still worked and continues to work. We know that we can use classical physics formulas to calculate the velocity and momentum of objects as they are moving on Earth at standard gravity. So maybe gravity was the key to understanding why objects in space behave differently? Bingo! The theory of general relativity is actually a gravitational theory, and once again, in true Einstein fashion, its genius is in its uncomplicated nature. The theory shows us how the gravitational forces of objects affect their relationship with each other and the space around them. That's it—nothing less and nothing more.

Classical physics measures gravity by looking at the mass of two objects and their acceleration as they move towards each other. This works well for macroscopic objects or things that we can see without the use of other devices. You can calculate acceleration due to gravity when you drop a book to the floor or a penny off of a roof. You can see these things as they interact on Earth, in standard gravity. But microscopic particles like molecules, atoms, and subatomic particles don't behave like this, either in gravity or in a vacuum, because of wave-particle duality, and the big things we see in space don't behave this way because of the absence of gravity. Therefore, classical physics can't be applied to explain the behavior of any of these objects.

General relativity aims to do just that, and Einstein's theory remains the only working hypothesis that is fully supported by empirical data.

The theory focuses on two simple premises- that gravity is just as relative a force as any other and that objects with a large gravitational mass have the capability to bend spacetime, which is the term used for the whole of the three dimensions plus time as the fourth dimension. The first part of the theory was based on Einstein's study of freefall. He postulated that an object falling on Earth and an object falling on a spacecraft would fall at the same rate if the spaceship were traveling at a speed correlating with the speed of gravity. Since gravity causes an object to fall at a rate of 9.8 meters per second, the same object would fall at the same rate in a spacecraft accelerating upward at the same speed. This proves that gravity itself is relative.

The second part of the theory, which focuses on spacetime, is actually a bit more complicated, but it makes a lot of sense once you grasp it. Think about space as a giant blanket, being held at all four corners and stretched as taut as it can be. There is linear space above and below the linear plane of the blanket. Now, place a ball in the center of the blanket. What happens? The mass of the ball forms a divot for it to sit in. And if you place a heavier ball onto the blanket, it will create a deeper divot. And if the second divot is deep enough to stretch out the fabric under the first divot, then the first ball will begin to roll towards the second one until they meet. Congratulations! You've just made a black hole!

Every large object in space is sitting in its own divot on the blanket of spacetime. Stars and planets of all sizes and masses, from dwarf planets to black holes, are all just trying to stay in their own divot and draw other things into their divots. It's that push-and-pull that keeps our solar system functioning. Think about the Sun, the largest object in our solar system. The closest planet to our star is Mercury, which is incredibly tiny compared to the Sun. You would think that Mercury would get sucked right into the Sun's spacetime divot, but it doesn't. Why? Because even though Mercury doesn't have enough mass to create a huge dent of its own, that lack of mass is what keeps it from "rolling" towards our star. It simply doesn't have the weight to *start* rolling. The Sun's gigantic mass and Mercury's tiny mass balance out with the distance between them and create a delicate gravitational dance.

Thus it goes with the rest of our solar system. Jupiter, our largest planet, is indeed massive, but it and the Sun are far enough apart to stay in their own divots. Smaller objects, like asteroids and comets, don't have enough mass to create their own dents, and so they rely on their gravitational ties with the Sun to stay in motion. Meanwhile, the planets also have their own satellites, like the Earth and its moon. The moon and the Earth are playing the same gravitational game that the Earth and the Sun are playing, and so on it goes throughout our solar system.

On a larger scale, we have gravity super-users like black holes. These objects aren't really holes but incredibly dense solid objects that have a powerful gravitational pull. A black hole has the capability of sucking in nearly everything that comes near it, including light, which why they appear to be the entrance to a void. The strongest black holes create a singularity, drawing in every bit of matter, electromagnetic wave, and energy until the very fabric of spacetime is pinched into a single point. Nothing within gravitational reach can escape these so-called monster black holes.

Of course, all this gravitational warping has an effect on spacetime. That's why the Kelly twins aged at different rates while one was in orbit and the other was on Earth. Even on the surface of our planet itself, you can see the influence of gravity on time. If you had two watches that were synchronized at sea level and took one to the bottom of Death Valley and one to the top of Mt. Everest, they would no longer be synchronized when you brought them back to sea level. The one that went to the desert would be slightly behind the one that climbed a mountain, where the effects of gravity aren't as strong, and the time passes more quickly. It's been calculated that if the clocks were left in those places over the course of a lifetime, the Mt. Everest clock would end up 40 hours or more ahead of the clock left in Death Valley. Time dilation is a very real phenomenon, on Earth and in space.

Einstein's theory of general relativity allowed for scientists in many disciplines to begin thinking about matter, mass, and gravity in a whole new way. Quantum chemists and physicists began to see atoms and molecules as tiny systems that relied on both gravitational and electrochemical bonds to stay together- which also gave them new ways to try and break them apart. Astrophysicists gained the capability to calculate gravity and spacetime with more accuracy, even on celestial bodies that are lightyears away, and began to reach new understandings about our own solar system. The theory of general relativity also gave physicists the ability to recognize and measure gravitational waves, which makes it easier to identify and measure large objects in space that are a great distance away.

Leaving Germany Behind for Good

Einstein's 1905 miracle year papers and the publishing of the general relativity theory in 1915 made him a bit of a rock star among the scientific community and beyond. In 1921, the physicist traveled to the United States for the first time, and he was enthralled with the vastness of the country and the attention he garnered from the American people. So enamored was he with the United States that he wrote an essay about his visit, praising Americans for being caring, self-confident, and optimistic. Einstein left America and toured through Asia and Europe, collecting accolades and delivering lectures on a variety of topics from physics to philosophy. In Palestine, he was greeted as a statesman rather than a scientist, and indeed, he was offered the first presidency of Israel when the Jewish state became autonomous in 1948.

A known pacifist, as we noted before, Einstein became increasingly concerned with the political climate in Germany, and the rest of Europe as Adolf Hitler rose to power in the early 1930s. Einstein was visiting the United States on a lecture tour and guest professorship when Hitler became the *de facto* dictator of Germany in 1933. Einstein's home near Berlin was pillaged, his belongings taken, his papers burned, and his property made part of a Hitler Youth Camp. The famed scientist, solely by the happenstance of being Jewish, could no longer return to the homeland that he'd long held such a love-hate relationship with.

Einstein immediately renounced his German citizenship, this time for good, and asked his powerful friends in England to help his Jewish colleagues escape the Nazi regime. Many of his Jewish compatriots had already been let go from their university and laboratory appointments and had gone into hiding, fearing for theirs and their families' lives. A relationship with Winston Churchill himself is what helped Einstein plead for his friends to be assisted out of Nazi Germany and into safety in the United Kingdom, where many institutions were happy to snap up the wealth of brilliance coming out of the German woodwork.

Despite all this, Einstein failed to secure his own citizenship in Great Britain, whose Parliament feared retribution from Germany should the scientist take up permanent residency there. Einstein then accepted an invitation from The Institute for Advanced Study at Princeton University in New Jersey to become their resident scholar. Yes, the world's foremost scientific mind decided he'd like to go live in New Jersey. What can we say? Genius takes many forms. By 1935, Einstein had become a permanent resident of the United States and gained his citizenship in 1940. Einstein remained an instructor at Princeton until his death in 1955.

Before cremation, Einstein's brain was removed by pathologist Dr. Thomas Harvey, without permission or sanction from the Einstein family. The brain was later sectioned into more than 150 pieces and preserved on slides for study. Eventually, Einstein's surviving family allowed for his brain to be used for research, under the caveat that it would only be used for science and not sensationalism. Upon his death in 2007, Dr. Harvey bequeathed the last remaining samples to the National Museum of Health and Medicine.

Albert Einstein made some of the most important contributions to modern science during his time as a student, teacher, and researcher, but that doesn't mean that he was without his human faults or wasn't controversial at times. In the next chapter, we'll examine parts of Einstein's life and work that wasn't always so sunny, and some things that he never quite got right, to get a better picture of the whole man, not just the wünderkind with the miracle year.

Chapter 5: Even Einstein Can't Always Be Right

By all rights, Albert Einstein was one of the greatest minds ever to grace Planet Earth, but he certainly wasn't without his detractors, and his personal life was a bit of a mess, too. Some of his colleagues weren't always the most supportive of his work, and although Einstein was no doubt brilliant, he didn't have the answer for everything. Let's go ahead and look at some of the things that Einstein didn't do so well with- because if the world's smartest scientist wasn't always perfect, then who could ever be expected to be?

Debating the State of Physics

Einstein and atomic modeler Niels Bohr considered themselves to be friends, but even friends don't always see eye-to-eye. When it came to the split between classical and quantum physics, the men were on different sides of the fence. Their disagreements led to a series of popular public debates on a variety of topics relating to physics, and Einstein wasn't always declared the winner. The first of these debates stemmed from Einstein and Bohr's professional disagreement over the theory of the photoelectric effect. Bohr was still a few years away from developing his atomic model and wasn't sure that Einstein's hypothesis made much sense.

In the realm of classical physics, it didn't. But Einstein was hooked on the idea of quanta because he thought it was an excellent way to marry the concept of wave-particle duality with something that could tangibly be measured. Bohr didn't see it that way, nor could Einstein understand why his friend was so unwilling to give up classical physics just yet. By 1925, however, when both scientists had been able to prove their respective sides through experimentation, they conceded that it was the very nature of quantum physics itself that had caused their earlier disagreement. So they moved on to arguing about other things.

The subject of their next debate, their first after the unofficial official split between classical and quantum physics, was about the future of quantum physics as a separate discipline. In the mid-1920s, fellow physicists Max Born and Werner Heisenberg (he of the uncertainty principle) postulated that they had learned everything they needed to know about quantum behavior and that the rest was up to probability. Einstein was incensed! He couldn't understand why his colleagues were so willing to give up on such a promising young discipline. Einstein openly stated that he didn't believe that God would leave the motion of even the tiniest particles up to chance and that he was going to continue seeking new answers in quantum mechanics.

Einstein was in for an even ruder awakening when Bohr decided to side with Born and Heisenberg. But Bohr not only sided with them, but he also went so far as to write a theory to reconcile the entirety of unknown quantum physics knowledge in one neat package. This theory, which Bohr called the principle of complementarity, said that all forces and objects in the universe have an equal but opposite force or object that keeps them in check. We're simplifying it here, but doesn't that sound awfully familiar? Like classical physics, maybe? Einstein was dissatisfied with Bohr's flippancy and with his theory.

Einstein also, as noted, didn't care much for Heisenberg's uncertainty principle. Einstein couldn't handle thinking that the universe was up to chance, nor that it had itself all sorted out through complementary pairs of forces and matter. It had to be one, the other, both, or something different, and Einstein wanted to be the one to find the answers. He wasn't content to let either of his friends be right, which eventually led to Einstein's work on Unified Field Theory, which we'll get to in a little while. Needless to say, no one was happy with the result of this debate, which raged over the course of 1927, and scientists today still reference the arguments of Bohr, Einstein, and Heisenberg during this time period.

In 1930, the men once again took up the debate about uncertainty, with Einstein relying upon one of his famous thought experiments. Bohr countered by building a physical working model of what Einstein described, and to nearly everyone's amazement, it showed that Einstein was wrong. (Do you think Bohr did a little happy dance? We like to think so.) The experiment and counterargument went like this:

Einstein performed a thought experiment that he was sure would prove Heisenberg's uncertainty principle to be inaccurate. In Einstein's scenario, he had a box full of photons, just doing photon things. Einstein surmised that the contents of the box could be measured and predicted by opening the box and letting some photons escape. His train of thought was that the remaining photons in the box could be measured by either number or total energy, and the difference would be the measurement of what left the box while it was open. Einstein's argument against the uncertainty principle was that accurate measurements could be taken using either time or energy as a variable. Therefore it wasn't that difficult to predict where the photons were going and would be, as the uncertainty principle suggests. (If you'll recall, uncertainty tells us that we can know where particles have been, but not where they are and where they're headed.)

Bohr was unhappy with Einstein's take, and so, like any rational, brilliant human being arguing with another genius, he built the box that Einstein described. The structure was suspended by a spring with a pointer attached to the side of the box corresponding with measuring devices mounted to the stanchion. Inside the box was a clock on a timer that would open a small shutter to let the photons escape at regular intervals and an electrode to produce the electromagnetic energy. Bohr argued that Einstein wasn't even taking into account his own theory of relativity in his thought experiment! If the electromagnetic waves were sending photons flying around the box, and the box was under gravitational force, then the photons' positions would depend on where they were being observed from.

Physical experiments conducted and measured using Bohr's model of Einstein's box proved to side in Bohr's favor. The photons demonstrated behavior described by the uncertainty principle AND were observed to move under the effect of gravity. If only Einstein had included "in a vacuum" in his part of the argument, the result of this debate might have been very different!

The last of Einstein and Bohr's public debates were centered around the theory of quantum entanglement. As you'll recall, this theory was born when Schrödinger was putting cats in a box with poison (a thought experiment, not in reality! He was a weird guy, but not inhuman.) Entanglement means that it's possible for particles and waves to get wrapped up in each other as they are traveling, leading to difficulties in gaining accurate measurements of where the particles are going. Five years after their debate over Einstein's Box, Bohr and Einstein found themselves at it again, and a theory that Einstein had written with two other scientists is what set it off.

The Einstein-Podolsky-Rosen theory, published in 1935, dealt with the behavior of waves and particles in their immediate environment- this was an offshoot of a lesser-known tenet of general relativity known as the principle of locality. The EPR paper, as it came to be known, addressed the ability of quantum mechanics to deal in what Einstein called "physical reality" and questioned whether the principles of quantum mechanics could allow for the real-time and real-space measurement of the position of particles that were entangled. In other words, if two particles are entangled, and it's possible to know where one of those particles is at a specified time, then it's impossible to know what the second particle is up to because the measurements would have to be taken faster than the speed of light. We know this is simply not possible, because according to Einstein's own theory of relativity, nothing can travel or act faster than the speed of light.

The EPR theory became a paradox when another possibility was introduced. What if you could measure the position of the particles using momentum? Again, if you know the speed of the first particle, you should theoretically be able to determine the speed of the second. Except you'd have to, you guessed it, be faster than the speed of light to complete the measurements that quickly. But if you know the momentum of a particle, you don't know its exact position. And round and round it goes. No one can determine where both particles are, where they are headed, and what their momentum is because these measurements can't be taken and calculated faster than the particles are moving. There is, according to EPR, no physical reality in which these calculations can take place.

Einstein, Podolsky, and Rosen were satisfied that they had come up with a theory that would stymy any arguments, but enter, of course, Niels Bohr. Bohr used his own hypothesis of the principle of complementarity to push back against Einstein's supposition that quantum mechanics was lacking in reality and was an incomplete science. Bohr desperately needed to show Einstein that the universe works in push-and-pull and that every particle, every wave, every invisible force, such as gravity, has a counterpart and that they're all interdependent on each other. Therefore, measuring particle A can absolutely give you information about particle B, no matter how fast they are moving or in what direction.

Einstein countered with a scientific "I don't give a rat's behind!" and tried to change the argument. Saying that Podolsky and Rosen had written the majority of the EPR paper regarding the measurement aspects of entangled particles, Einstein said that he himself was only interested in the local and non-local aspects of the argument. His new position was this: If particle A is measured using either time, energy, or momentum, then particle B is already off into the ether, doing its thing- so how could you know what particle B is doing? It's already out of its original locality, and so is the observer! Einstein also argued that if you measured particle A using time, it would have a different effect on particle B than if you measured A using momentum. Therefore, Einstein posited, particle B could exist in a multiple number of positions, but you would never know which position was the "reality."

Yeah, we know- this is a challenging paradox to wrap your head around. It was so ridiculously complex that no one was declared the winner of this particular debate, and it was pushed to the side for YEARS because no other scientists wanted to tackle it. It wasn't until 1964, nearly three decades later, that a physicist named John Bell proposed a theory of non-locality that seemed to solve the EPR paradox. Bell, using theories on hidden variables, developed since the 1935 debate and was able to prove that the problems lie in the locality and non-locality of measurements, not in the nature of the entangled particle themselves. How you measure the particles' behavior has little to do with what happens after- the true nature of quantum mechanics lies in its non-locality. What one observer sees from their locality is not what another will see in theirs unless they measure from the exact same plane at the exact same time. The debate rages on to this day as to whether or not that is physically possible.

So, if you were keeping score, Einstein won the first debate about the photoelectric effect. Bohr decisively won the second about the uncertainty principle. And even now, no one knows who was correct about the EPR paradox, and scientists are still arguing about it decades later. Such is, perhaps, the very core of quantum mechanics. If thinking about this stuff gives you a headache, be assured you aren't alone. Physicists have been devoting their lives to studying the behavior of waves and particles for well over a century now, and one of the only things they know for certain is that they don't know everything. That's what makes it so fascinating. What we do know is that wave-particle duality, entanglement, and the uncertainty principle are, and will continue to be, a source of great debate.

Outside the Classroom

Albert Einstein is hailed as one of the brightest minds in human history, but we often forget that he was just that- human. He wasn't a perfect scientist, as demonstrated by some of his failures and losing arguments. He wasn't a perfect friend or colleague, as we see when he threw Podolsky and Rosen under the bus to try and win the last debate with Bohr. He definitely wasn't the best husband and father, either. Remember when we referred to Mileva as his first wife in the last chapter? And said they had started a young family while finishing school and entering into the Patent Office years? Yeah, that didn't go so well, to be honest.

For starters, there is the mystery of what happened to Albert and Mileva's first child, a girl named Lieserl, who was born in 1902, nearly a full year before the Einsteins got married. Mileva allegedly went home to family in Serbia to finish out her pregnancy, and then the baby just disappeared, for lack of a better term. The official word was that Lieserl died of scarlet fever in infancy before Mileva could return to Albert's side in Switzerland, but it is more likely that the baby was put up for adoption because her parents were unmarried.

From there, the story of Albert Einstein's personal life becomes even more convoluted. He and Mileva stayed together out of mutual admiration for each other's intelligence and had what, from the outside, looked like a typical, albeit stiff, union. They had two sons together, Hans Albert, born in 1904, and Eduard, born in 1910. As

Albert moved the family around to take different teaching positions, Mileva grew increasingly dissatisfied with their life, wanting her boys to have a more stable upbringing. By 1914 and the start of the First World War, she'd had enough. She decided to separate from Albert, who didn't put up too much of a fight; he even promised her money from any future Nobel Prizes to support her and the children. He made good on that promise in 1922.

Part of the reason Mileva was upset is that Albert wasn't exactly the most faithful of guys, and she knew he was corresponding with other women while on his teaching travels. One of these women was Marie Winteler, the daughter of family friends with whom Albert had boarded while he was attending school in Switzerland as a youth. Another was Elsa Lowenthal, née Einstein- a cousin both maternally and paternally, and someone who Albert was close to his entire life. Elsa divorced her first husband, Max Lowenthal, in 1908, and by 1912, was having a full-blown affair with her married cousin Albert.

When Albert and Mileva finally divorced in 1919, a full five years after becoming separated, Albert and Elsa married within six months. He raised Elsa's daughters as his own and claimed a closer relationship to them than his biological sons with Mileva. To his credit, he often lamented this fact and took responsibility for his absenteeism, especially after Eduard was diagnosed with schizophrenia and sent to live his adult life in a sanitarium. Elsa remained with Einstein until she died in 1936. She acted as his conscious, his gatekeeper, and his confidant while he immersed himself in his studies and teachings, and yes, his other affairs.

Even while Elsa did everything she could to protect Albert's personal life, it was well-known that he was carrying on elsewhere. He had no fewer than six extramarital dalliances while he and Elsa were married and took up with much younger women after her death. He didn't discriminate or have a type; his partners in philandering were women of all ages, from all across Europe and the United States, and from all social classes. At the time of his passing in 1955, he was engaged in a relationship with a Russian-born librarian who was 22 years his junior. Elsa, for her part, was philosophical about their marriage and Albert's inability to stay faithful. Paraphrasing a letter that the second Mrs. Einstein wrote to a friend, she said that nature had made Albert a scientific genius, which meant that he had to be deficient in some other way- which happened to be infidelity.

In other weird Einstein news, he also had a strange obsession with plumbing; so much so that when he lamented that his theories had been used in the development of nuclear weapons, he said that he wished he'd been a plumber or a peddler rather than a scientist. The Plumbers and Steamfitters Union granted that wish with an honorary membership in 1954. Einstein also never outgrew his love of music and was an avid concertgoer and violinist throughout his adult life. He had the opportunity to play with many popular orchestras and string quartets as a guest performer. He was even erroneously credited as the curator of an important collection of Mozart's works. That distinction actually belonged to distant cousin Alfred Einstein, who also emigrated to the United States in the 1930s to escape the Nazi regime.

Politics played an increasing role for Einstein as he aged. He embraced his Judaism after leaving Germany for good and became heavily involved in the issues surrounding Palestine and the quest for an independent Jewish nation. When Israel became an independent country in 1948, Einstein was offered the opportunity to become the fledgling nation's first leader, which he politely declined. He would, however, continue lecturing on human rights, religious freedom, and other philosophies and was a founder of Hebrew University in Jerusalem. Einstein greatly admired the passion for nonviolence demonstrated by Mahatma Gandhi and sat on the board of several human ethics organizations. Isn't it strange, though? It's difficult to reconcile Einstein the scientist, Einstein the hobbyist, Einstein the terrible husband and father, and Einstein the human rights advocate. All these people are the same man, who was as complicated as he was brilliant.

Back to Science, Sort Of

One thing you might have noticed about this discussion of Einstein after emigrating to the United States is that we haven't talked all that much about actual science. Aside from his debates with Bohr, Einstein wasn't really working on anything new but instead trying to reconcile some of his old theories with more contemporary developments, but to little avail. He honestly wasn't all that interested in work that wasn't his own. It's a major criticism that Einstein didn't produce much in his later years, but to be fair, relativity was pretty huge- that's a lifetime achievement, so who cares if he rested on his laurels a bit? He sort of deserved some rest, from a scientific standpoint.

Einstein's contributions as he aged were in theoretical physics rather than laboratory science. One factor in this was, believe it or not, the rise of the popularity of science fiction as a genre for literature and film. People wanted to know if time travel was genuinely possible and if great wormholes in space did exist, and Einstein wanted to consider these prospects, too. On the subject of wormholes, he teamed up with friend and colleague Nathan Rosen (the R in the EPR paradox) to try and figure out if wormholes could potentially be a real thing.

The theory, that Einstein and Rosen devised, manipulated two black holes to act as the positively and negatively charged ends of a cosmic tunnel or wormhole. Through this wormhole, matter could travel from Point A to Point B in a rift in spacetime. Such theoretical wormholes were dubbed Einstein-Rosen bridges, but sadly, they don't exist, at least not in the scope of our current knowledge. In order for wormholes to be functional tunnels through spacetime, there would have to be a lack of gravitational collapse. An Einstein-Rosen Bridge might allow a particle to pass through the event horizon (the black hole's point of no return) and might even allow that particle to travel the length of the wormhole and emerge on the other side, but it wouldn't be able to make the return trip. And that's assuming that the particle's energy didn't disrupt the gravitational field of the wormhole enough to cause it to collapse. There are so many variables at play in an Einstein-Rosen Bridge that even the mathematical likelihood of them existing doesn't speak to the reality of such a phenomenon.

Einstein's research into wormholes between black holes did open the doors for other theoretical physicists to take up the search for truly traversable wormholes in the fabric of spacetime. Secondary to his experiments into wormholes were the Einstein field equations, which help mathematically describe some of the functions of general relativity. These equations are of great import to astrophysicists and cosmologists as they plumb the depths of outer space and strive to explain the behavior of celestial bodies that are light-years away.

One of the last things Einstein worked on in his lifetime was trying to create a Unified Field Theory, or again, the theory of everything. He published his papers in 1950, just a few short years before his death. Sadly, Einstein himself missed a few critical points in his research (like pretty much ignoring strong and weak bonds among atomic particles), and the papers were not well-received by the scientific community. In his quest to be the first to connect all of quantum theory into a bundle as neat as, well, a quantum, he managed to alienate himself from whatever colleagues and friends he had left. It's sort of a sad end to the life of a genius. Einstein's last few years were spent in relative isolation from the rest of the world, but his legacy remains in the groundbreaking theories he produced and proved as a young man.

A real problem with the unified theories that Einstein was trying to reconcile in his last years was that he wasn't paying attention to the developments that were happening around him. He was too mired in his past achievements and preconceived ideas to be willing to accept that the field of physics was marching quickly towards the future without him. His pacifism and refusal to work on any sort of nuclear power plants or nuclear weapons also precluded him from getting any hands-on experience working with those applications of quantum physics. Einstein was brilliant, yes, but as his second wife said, lacking in areas that weren't directly related to the application of science. Perhaps hubris and humility were also among his deficits.

But enough about Einstein- we've already given him two whole chapters! And sure, we'll probably mention him a few more times before we're done, but let's take a look at some of those 20th-century advancements that Einstein wasn't paying attention to when he wrote his UFT. Maybe you'll get some inspiration to write a Unified Field Theory of your own. Or build a nuclear power plant. Either way, let's go learn some more stuff!

Chapter 6: Rapid Advances in the Mid-20th Century

Welcome to the next chapter, where we're going to keep on plowing through the 20th century. The significant discoveries and groundbreaking theories that came from the 1890s through the early 1930s set the table for the split from classical physics, and the next few decades were a whirlwind of research and development in the fields of quantum physics, theoretical physics, quantum mechanics, and quantum chemistry. Sometimes the lines between these disciplines get a little blurred, but they are all based on the principles of quantum theory, and the math, science, and innovations from this mid-century boom continue to be applied across the world today. Let's get back to our physics timeline with a little bit of math, magnetics, and medicine.

Don't Dirac the Boat

While Einstein and Bohr were busy with their debates in the late 1920s into the 1930s, English physicist Paul Dirac was hard at work in his laboratory trying to figure out what was really going on inside an atom. When we say hard at work, we mean it. Dirac was a fiercely dedicated scientist, a man of so few words that it became a subject of teasing by his peers, and remarkably socially awkward before socially awkward was a thing.

Looking back, it's entirely possible that Dirac was not neurotypical, although autism was not an official medical diagnosis until the 1940s, and even then, the spectrum wasn't introduced until the 1990s. But we digress- suffice it to say that Dirac was definitely not the life of the party, although he did have a circle of close friends and married Margit Wigner, the sister of fellow physicist Eugene Wigner. He was known to be an exacting man, much to the chagrin of his colleagues, who learned to watch everything they said around Dirac, lest he chastised them for imprecise use of language.

Dirac was fascinated with the inner workings of atomic particles. His two biggest contributions to science (and there were many, but we'd run out of time and words) were the marriage of relativity and quantum mechanics to describe the behavior of the electron, and the Dirac equation, which explains quantum field theory based on gravitational fields. There are even bonus calculations if it were determined that all of space had a single monopole magnetic field governing its behavior. Okay, that's a lot to chew on, but Dirac's equation also implied something crucial to furthering theoretical physics and astrophysics, and that was the existence of anti-matter. (!)

As for bringing together relativity and quantum mechanics, that was huge as well! It's almost like other physicists had been arguing about how to do that...oh, wait. Well, while everyone else was debating, Dirac was doing. He released his quantum theory of the electron in 1928, but the seeds were planted while Dirac was working on his Ph.D. a few years before and stumbled across some unpublished work by Heisenberg, who would become one of Dirac's few lifelong friends. Heisenberg's paper piqued Dirac's interest because of how it spoke about the relationship between particles and position (perhaps a precursor to the uncertainty principle?), but it lacked in the one thing Dirac loved more than anything-mathematics.

So, if Dirac published several viable versions of a quantum theory in the mid-to-late 1920s, with math to back them up, why didn't anyone pay attention? For starters, Schrödinger had also just released his theory of wave mechanics, and Schrödinger and Heisenberg were really busy hating each other over it. Einstein and Bohr were all wrapped up in their own much friendlier debates. Dirac's theory was almost ridiculous in its brilliance and its (relative) simplicity. He looked at classical Newtonian and Hamiltonian physics and observed that they were using bracketed equations to measure the evolution of motion over time. Dirac adapted those bracketed equations to the movement of particles, accounting for relativity and time dilation, and voilá, a quantum theory that married quantum mechanics with the reality of observing and tracking particle movement. That's what everyone had been looking for, right?

In his quantum theory and quantum theory of the electron, along with his equations, Dirac managed to satisfy many requirements for discovering the true nature of quantum movement. He even went so far as to correlate the classical physics laws of conservation of matter and energy and propose that electrons and photons could be swapped for one another in equal exchange. That is to say, where a photon disappears, an electron appears, and where an electron disappears, a spark of light (a photon) appears. In his quantum theory of the electron, Dirac even worked this out to account for the potential for anti-matter- what happens when an electron and another particle meet and cancel each other out in a burst of mass-annihilating energy. Absolutely wild stuff- and the basis for atomic fission science later in the century.

Dirac was also obsessed with putting everything into order, and operations were a huge part of his work. He used operations in many of his mathematical premises because he wanted everything he did to be performed with accuracy, not only in his own work but so that others could recreate his results (basic scientific method stuff right there.) Think about when you were in grade school and learned about Please Excuse My Dear Aunt Sally – parentheses, exponents, multiplication, division, addition, and subtraction. This basic order of operations lets you know how to perform an equation so that you get an accurate answer. Order of operations also tells us how to put the condiments on the *inside* of our sandwiches and put our socks on before our shoes. Anything done in the incorrect order gives us less-than-desirable results.

These bracketed, organized, ordered operations are what gave life to all of Dirac's theories and equations. Over time, other physicists and mathematicians began to realize the importance of Dirac's work and gave it the attention it deserved. While Dirac continued to spend his life devoted to particle theory and relating it to mathematics, he is also remembered for his quirky personality, his non-concern with what others thought of him, and his devotion to translating the written word of physics into viable, well-ordered equations that would become the universal language for physicists everywhere. Bonus for those who like a good headache- here's what Dirac's equation of quantum theory of the electron looks like:

The Dirac equation in the form originally proposed by Dirac is:

$$\left(\beta mc^2 + \sum_{k=1}^{3} \alpha_k p_k c \right) \psi(\mathbf{x},t) = i\hbar \frac{\partial \psi(\mathbf{x},t)}{\partial t}$$

Have fun! Dirac received the Nobel Prize for his work in 1933 (that's right- the same year as Schrödinger! Even the prize committee couldn't decide whose theory was correct.) That's when his fellow physicists sat back and took notice. Even Einstein had to concede that the Englishman was a rare genius.

Picking Back Up with Particle Theory

If you'll recall, the discovery of the atom and accurate atomic models developed over time, as did the revelation that the atom itself was made up of a lot of tiny, constantly moving parts and that those particles also moved in waves. As physicists continued to study these tiny particles, originally thought to be the smallest building blocks of matter, it became clear to them that atomic particles themselves were made up of even smaller sub-atomic pieces.

Those pieces remain the centerpiece of quantum mechanics and physics today, and new ones continue to be hypothesized and discovered.

From the earliest atomic theories, we gained the knowledge that atoms are made up of a core, called the nucleus, and the electrons that orbit it in a delicate dance. We also know that there are positively-charged particles in the nucleus called protons, and that electrons are negatively charged, and that basic electromagnetism is what holds the atom together. Our friend James Chadwick, who you'll remember managed to conduct experiments while a prisoner of war, contributed to the discovery of the neutron to the conversation in 1932. It was also around this time that other physicists were starting to take Dirac's work into serious consideration, leading to the first proven observation of the positron (or anti-electron). These were the particles that Dirac posited would annihilate one another should they collide. Anti-matter is real, folks!

Keeping in mind that this is a book for beginners, and there is no quiz at the end, let's run through some of the major sub-atomic particles, their classifications, and relationships. Just being familiar with the names and simplified functions will help you have a greater understanding of the tiniest inner workings of quantum mechanics.

The first particle we'll look at is known as the "fermion," named for quantum electromechanist Enrico Fermi, who was interested in many of the same concepts as Paul Dirac. He and Dirac both mathematically predicted the existence of fermions independently of each other at around the same time in 1926, and Dirac always gave the credit to Fermi, although the official name for the equations is the Fermi-Dirac statistics. What their calculations proved is that all particles have a set angular momentum, which causes them to spin, and that particles can be classified by exactly *how* they spin. Particles that are classified as fermions spin with an angular momentum that falls on a half-integer (i.e., $n + ½$.)

Fermions, as discovered over the decades since, can be split into a few categories of their own. Hang on tight because we're going to try to wrap this up as neatly as possible! There are two types of fermions- quarks and leptons. Quarks are the subatomic particles that make up most of the matter we are familiar with. Quarks come in six "flavors"- up, down, charm, strange, top, and bottom- and it is quarks that combine to make larger particles known as hadrons. Hadrons help up drift back into more familiar territory. Hadrons made of quarks are what make up the two of the parts of the atom that we're comfortable with- the protons and neutrons.

Quarks become hadrons due to the strong fundamental force of physics. (Remember- strong, weak, electromagnetic, gravitational… we talked about these forces a while ago.) So, quarks make up hadrons, hadrons make up atomic particles, atomic particles make up atoms, and so on through molecules and compounds until we have recognizable material. Quarks also carry the polarity within an atom, and when they combine in odd numbers to create a positive or negative charge, the particle they create is called a baryon. A baryon is also classified as a hadron because it is made up of quarks. Each quark makes up a set fraction of an atomic particle, and each positively-charged quark has a negatively-charged counterpart called, creatively, an antiquark.

The other type of fermions, called leptons, are classified thus because they interact under the weak fundamental force of physics and because they spin at full integer speeds, not in halves. There are six types of leptons, the most familiar being the electron. The other leptons are the electron neutrino, the muon and muon neutrino, and the tau and tau neutrino. Electrons, muons, and taus are negatively charged, and the neutrinos are neutral, as their name would imply. Together with quarks, leptons form the building blocks of the atom and eventually all recognizable matter. Quick recap: Fermions are split into quarks and leptons. There are six kinds of quarks and six kinds of leptons, and when they perform their respective functions, they make up the atoms that make up most matter. Got it? Great job!

The next type of classified subatomic particles is the bosons. Bosons are the sub-atomic particles responsible for creating the fundamental forces within an atom, and there are five experimentally-proven elementary bosons. These are the photon, the W boson, the Z boson, the gluon, and the Higgs boson (which was finally seen in 2012!). Theoretically, there is a sixth unproven boson called the graviton, and this is the boson that physicists hypothesize is responsible for gravitational forces. It's yet to be observed. Where we're from, the graviton is a centrifugal carnival ride responsible for a lot of regurgitated fair food...or was that the Gravitron? Regardless. *Shudder*

Anyway, bosons are tasked with keeping things together, as it were. They behave in a way that causes the fermions to find each other and begin producing the forces necessary to become an atom and eventually a measurable quantity of matter. Bosons also behave in correlation with the Bose-Einstein statistics, rather than the Fermi-Dirac statistics, which means that bosons do not act the same under a change in conditions as fermions do. Fermions change state based on environmental fluctuations, and bosons change their behavior based on energy fluctuations. When two bosons connect, usually a charged boson and a neutral boson, they form a compound particle known as a meson. Is your head spinning yet? No? We must not be trying hard enough.

What you really, absolutely need to know about sub-atomic particles is that they are still being explored and described with greater detail all the time. Scientists are always seeking new ways to classify these particles, and every time they get a new piece of information, it brings us closer to knowing how each specific particle works together to create matter and energy. The closer we get to understanding it, the closer we get to utilizing that information for our benefit. *That's* the curiosity and the motivation that keeps particle physicists going.

Nuclear Advancements in War and Peace

The power of the atom, once unlocked, is undeniably a force to be reckoned with. Once Einstein presented his mass-equivalency equation to the world, the race was on to see who could be the first to use the energy trapped inside an atom. What would follow were major developments that changed the world forever, for better or for worse. The premise is simple- split an atom, and channel the resulting energy release. This is what is behind nuclear engines, nuclear power plants, and yes, even a nuclear bomb.

Some materials are better suited to nuclear applications; these are most commonly radioactive elements with reasonably long half-lives. All elements will eventually break down, but the most radioactive elements- radium, uranium, plutonium, and the like- have specific radioactive properties that make them especially useful for these purposes. Because their atomic bonds are already weak, breaking them apart requires less effort, and they can often be broken down with a simple bombardment of neutrons.

These theories were first tested in laboratories in the 1930s, in a series of experiments conducted by German and Austrian scientists Otto Hahn, Lise Meitner, Otto Frisch, and Fritz Strassman. The researchers conducted trials with uranium, attempting to split the very nucleus of the atom. They were successful in doing so in 1938, and the nuclear era was born. The uranium acted precisely how the physicists had hoped, and they dubbed their new process nuclear fission.

A year later, Hungarian physicist Leo Szilard, who had also been working on similar experiments, presented research to Einstein, asking for him to review and sign off on it. That research? It proved that if a uranium atom were correctly split, it would emit just the right number of electrons to continue a chain reaction of atomic breakdown, releasing energy at every fission. Once signed by Einstein, that research became the basis of the Manhattan Project, the American program to develop the atomic bomb. Many of Einstein's familiar colleagues were counted among the Manhattan Project Scientists, including Bohr, Fermi, Chadwick, Wigner, and Oppenheimer, just to name a few. The project was conducted at numerous sites across North America under great secrecy.

Einstein himself never worked on the design; in fact, as a pacifist, he opposed its development, and the United States government chose to keep him away from the project lest he decided to sabotage it. The United States wasn't the only country working on the development of atomic weapons, but they were the first and only to successfully (if you could call it that) deploy a nuclear weapon during armed conflict. The scientists themselves warned US President Harry S. Truman *not* to use the bombs on Hiroshima and Nagasaki, advising that their power and lingering radioactivity would cause needless suffering, but Truman ignored their plea, saying that they'd been paid to do a job and that it would be a waste of time and money and a sign of weakness not to use the weapons. Today, several nations have nuclear capability, but the specter of what happened in Japan at the end of WWII was enough to be a factor in the Cold War but also prevent global nuclear warfare.

Nuclear power plants take a gentler approach to harvesting the power of radioactive decay, but as anyone who lived through the 1970s and 1980s knows, it wasn't always the safest form of clean energy. If you've ever seen a nuclear power plant, you know they are almost universally distinctive because of their immense cooling towers. These towers often look like massive, steaming flower vases rising from the surrounding landscapes, and they are a necessary component to any nuclear power facility. The structure of a nuclear power plant is designed to contain any stray radioactivity and provide an isolated environment for the radioactive fuel, which in the majority of commercial nuclear power plants is uranium.

There are two main styles of nuclear reactors, the boiling water reactor (BWR) and the pressure water reactor (PWR). About two-thirds of the world's nuclear power plants are PWRs, but the concept behind both types is the same. Within a large concrete structure is a holding tank that contains fuel rods made from pellets of power-grade uranium. Water is pumped through the holding tank and is heated by the fission reactions occurring as the uranium breaks down and releases energy. The heated water and steam are pumped out of the tank where the steam powers a turbine which produces electricity. The remaining water flows through the cooling tower and is distilled back into a tank to start the process all over again. It's a simple, effective, carbon-free way of producing electric power. There are, of course, repercussions of this type of energy production.

In 1986, the Chernobyl nuclear disaster in Pripyat, Ukraine, caused the instant death of two plant workers and dozens of more fatalities in the immediate nuclear fallout. An electrical test of a systems switch went very, very wrong, resulting in the meltdown of two reactor cores that could not be adequately cooled. Radioactive steam blew out the side of a containment building, and the subsequent damage left the area uninhabitable. In 2007, a massive earthquake and following tsunami in northern Japan caused a meltdown at the Fukushima Daichi Nuclear Facility, where three units had a complete failure and others had hydrogen explosions from superheated water.

There is also the problem of what to do with spent nuclear fuel, which continues to be radioactive long after it has lost its utility. Most facilities keep their old fuel rods on-site, in deep cooling pools in concrete and boron-lined buildings. This is to prevent their continued radioactivity from leaking and keep their temperature under control. As the world uses more and more nuclear power, these rods will keep piling up until a more permanent solution can be found.

Many of the world's navies now use nuclear propulsion for their vessels, especially submarines, and this works well because there is always a water source for fuel-rod coolant as they travel the oceans. Some civilian ships, albeit those with commercial or research purposes, have also begun to use nuclear propulsion. It will likely be a long time before anyone can just go and buy a nuclear-powered pleasure boat- it's doubtful that we Joe Shmoes will be allowed to have fuel-grade uranium just docked at the marina anytime soon.

Harnessing Physics for Healing

Since the discovery of x-rays in 1895, this form of electromagnetic radiation has been used in the diagnosis of injuries and illness. Marie Curie herself was responsible for putting hundreds of mobile x-ray units into battlefield ambulances in World War I. X-rays are just one of the many applications of quantum physics at work in health care, and over the decades since they were first put into use, many other advancements have harnessed the power of wave-particle duality and the electromagnetic spectrum to heal the human body.

If you've ever had a soft tissue injury, maybe you had an MRI to help diagnose the extent of the damage. Magnetic resonance imaging is a direct application of quantum mechanics, and it uses nuclear magnetism to create an invisible field that penetrates the body's tissues and produce an image that doctors can use to determine treatment options. Perhaps you've had a CT scan or a PET scan. These electromagnetic wave diagnostics produce accurate cross-section graphics of tissue, giving medical professionals information that once could only be obtained by exploratory surgery. Quantum physics has saved a lot of people from unnecessary procedures.

Quantum chemistry aids in diagnostics too, which you'll know if you've ever had to have a contrast test that required a barium swallow.

Lasers are also frequently used in surgical applications these days, and we can thank quantum physics for that, too. The ability to safely control beams of condensed electromagnetic energy has made a world of difference in procedures such as corrective vision surgery, nerve ablations for pain management, and dissolution of foreign growths like kidney stones, gall stones, and fibroid tissue. It should be noted that lasers are also widely used in cosmetic procedures like hair removal and for the amusement of house cats. We doubt the developers of the first laser applications saw that one coming.

One of the most interesting and long-standing applications of quantum physics and quantum chemistry in the healthcare field is the use of radiation treatments for cancer. Just as x-rays were put to use in medicine within a few short years of their discovery, so too was radiation in other forms. It didn't take long after Becquerel, and the Curies discovered radioactivity that they also realized that it could both cause and cure cancer. Early attempts at using radium to shrink tumors proved to be rudimentary, and many of the caregivers developed blood cancer (much like Marie Curie's leukemia) in the pursuit of finding appropriate dosages for their patients. In the century since, radiation therapy has evolved into a reasonably safe, pretty darn effective proposition. Targeted radiation can be applied directly to the locale of the cancerous cells, dosages and exposures are calculated with extreme precision, and the side effects have been minimalized to the best of science's current abilities.

Wow, that was a ton of information about the inner workings of the atom and practical advancements in quantum physics! Sorry if it was a bit overwhelming, but quantum applications really do affect everything we do because, by its nature, quantum studies are all about understanding and harnessing the behavior of everything around us. In the next chapter, we'll examine some of the tools scientists use to develop their theories and observe their experiments. When you're studying things that can never be seen with the naked eye, it takes a lot of equipment to make things happen!

Chapter 7: Building Big Things...for Science!

One of the coolest things about the study of quantum physics is the equipment that goes along with the discipline. Because quantum physicists perform work with the tiniest sub-atomic particles and theorize about some of the biggest celestial bodies in the universe, it takes a lot of specialized instruments to complete their experiments and write their hypotheses. Let's take a look at how the tools evolved alongside the trade and how they employ some of the technology developed along the way.

Observing the Tiniest Objects

Remember our friend Leo Szilard, who convinced Einstein to sign off on the research that launched the Manhattan Project? It turns out that's not his only claim on quantum history. Szilard was also an inventor, and in 1928, he submitted the first known patent for the electron microscope. (He also filed patents for two early particle accelerators; this guy was incredibly ahead of his time!) Because Szilard had to flee Nazi occupation and hightail it to the United States like many of his other colleagues, he never had the opportunity to see many of his inventions come to fruition, but others used his concepts to build their own.

The scanning electron microscope, as we understand it today, is used across a broad spectrum of scientific fields, and it uses the power of the electron to define images and provide feedback that gives the observer an accurate picture of what they are trying to study. In a boring regular microscope, visible light is used to illuminate the material on the specimen slide, giving the researcher a good idea of what they are looking at, but not the greatest. With an electron microscope, a concentrated beam of electromagnetic energy is used to interact with the specimen material, highlighting every nook and cranny of the slide. The images produced are quite detailed and remarkable, and computer technology means that resolutions can be adjusted to higher magnifications than ever before. Contrasts and colorations can be added for even better results. Scientists of all disciplines appreciate the development of electron microscopy. You can even use an advanced electron microscope to see other electrons!

Another application of quantum physics that makes studying quantum physics easier is the mass spectrometer. You may think of this as a thoroughly modern machine used in television crime lab procedurals, but the mass spectrometer can trace its roots all the way back to 1896. While today's equipment for mass spectrometry is technologically advanced, the concept stays the same; there are three components that make up the machines- a source of ionization (the artificial charging of known particles to create an electromagnetic field), a mass analyzer, and a detector. Together, these work to tell scientists the atomic or molecular makeup of an unknown substance. This is tremendously useful in identifying elements and is utilized in forensic science, chemistry, biology, and, yes, quantum physics.

The earliest mass spectrometers used rudimentary electromagnets to produce ionization, and modern versions are still used today. Other methods of ionization include gas induction, electron ionization, and plasma induction. No matter what type of ionization is introduced, the intent is the same. These charged particles bombard the material sample, and the detector measures the feedback from how the particles interact with the sample. The results show the scientist the properties and makeup of the sample based on how it reacted to the ion bombardment. It sounds complicated, but it's another example of brilliance in its simplicity. It's also diagnostic science in action. The more researchers use mass spectrometry, the larger their database of known substances, and that's exceptional information for the furthering of human knowledge.

Observing the Farthest Objects

From the very beginning, electromagnetic waves have played an enormous role in the development of quantum physics, and the advancement of the understanding of these waves led to some really big things. Or we should say, led to *the discovery* of some really big things. Or both. When you think of a telescope, you might think of the humble mirror tube. It refracts light to magnify objects that are far away. That's classical physics in action, and it should never be discounted. Most of the objects within our solar system were discovered with this type of telescope, long before quantum physics was a thing.

But the introduction of radio waves for practical applications like communications got early quantum physicists to thinking, what if we could send radio waves out into space and see what happened? Could we begin to see and hear things we couldn't see and hear before? The answer is a resounding yes. One of the beautiful things about radio telescopes is that they can be used in daylight, whereas optical telescopes require darkness, so they can take in light only from the source they are trying to observe. Radio telescopes operate on a simple premise- send electromagnetic waves out into the universe and see what comes back to you. Since we know that all objects emit radioactivity in some form, how that radioactivity interacts with the known wavelength of the radio telescope gives scientists the data they need to know what's going on out in space.

For instance, radio telescopes can "see" celestial bodies because the waves sent out by the telescope will distort around anything in their path. Imagine you are holding a rubber ball under a faucet. The water doesn't go through the ball; it bends around it. You can still see the solid outline of the ball itself, and that's how radio waves act when they meet a solid object in space. Scientists can use the behavior of the waves to "see" the outline of faraway stars and planets. Amazing, right? Radio telescopes can also be used to send long-range communications, as in some of the first crewed space missions that used these telescopes to talk to the people back on Earth. And, of course, there is always SETI, which is determined to sit around listening for signs of life from other planets. But honestly, if you were from an advanced alien society, would you really want to talk to us?

It's a combination of classical and quantum physics that drives human space exploration, too. We need classical physics to help us get objects and people out of Earth's atmosphere and away from the planet's gravity, but once in space, quantum physics is needed to calculate time dilation, gravitational waves and forces, and the behavior of extraterrestrial material and cosmic rays. It's a group effort to put things from Earth into space, and classical physicists, quantum physicists, planetary scientists, engineers, chemists, and people from innumerable other disciplines all put their best feet forward to make space exploration a reality.

Manipulating the Atom

If you stop to think about it, quantum physics is intrinsically a discipline that speaks volumes about the human condition. It is driven by intense curiosity and our inability to say, "enough is enough." Quantum physicists are continually striving to find the next thing and the next, and as soon as they discover something, they want to know how they can use it for their own purposes. It might sound like we're trying to paint these researchers in a poor or selfish light, but that's not true. We're just pointing out that quantum physicists are all of us, eager to play with new toys; they just have way cooler toys.

Almost as soon as scientists knew what an atom was, they wanted to figure out how it worked. When they knew how it worked, they began to want to tear it apart and put it back together, much like a mechanic with a neat engine that needs repair. Along with Einstein's mass-equivalency equation came the knowledge that mass could be turned into energy and vice versa, as demonstrated with the development of nuclear weapons and nuclear power. But mass-equivalency also gave scientists a *big* idea- what if they could build a machine that would allow them to break apart and put back together atoms so they could observe the behavior of particles all the time? The ideas for the first particle accelerators were born.

The world's most famous particle accelerator resides at CERN, a massive research facility located just north of Geneva, Switzerland, and governed by the European Organization for Nuclear Research. A particle accelerator is a device that uses concentrated energy to create focused beams of particles that travel at high speeds and energies. Why would you want to do this? By accelerating particles up to these speeds, scientists are able to observe their behavior and extrapolate data about atomic breakdown as the material approaches the speed of light. The thing about particle accelerators, though- the smaller the object you want to observe, the larger your accelerator needs to be to get those objects up to speed.

The Large Hadron Collider at CERN is built in a ring that is nearly 17 miles around and is built underground to be able to contain all the accompanying machinery, such as cooling mechanisms and the power cells for the magnets that make the collider functional. This super-vacuum tube can accelerate particles towards each other in the ring at high energy, and when they collide, the atomic energy can be measured, and the tiniest sub-atomic particles can be observed. It was at CERN's Large Hadron Collider that the existence of the Higgs boson was confirmed in 2012. If you think that the size of the LHC is insane, wait until you read this. Its successor is being planned, and it is going to be a whopping 100 km around- that's a whopping 62 miles, and it will span the border between Switzerland and France. This new collider would have seven times the capability of the current Large Hadron Collider. Plans may also be in the works for a new miles-long linear accelerator too. All in the name of science! Research done in these large-scale accelerators gives us crucial data about the smallest building blocks of matter, but the sheer size of this equipment can be mind-boggling.

All the diagnostic and research equipment we talked about in this chapter was both made possible by and for the benefit of quantum studies. Pretty cool, huh? It's a self-sustaining discipline, which we think is one of the best things about quantum physics. There's always more to explore, whether it's going smaller or going bigger, and the future is wide open. In our last chapter, we'll look to the late 20th and early 21st centuries, in with some current affairs, and see what the next century might bring to the universe of quantum physics.

Chapter 8: What's Next? Quantum Physics in the 21st Century and Beyond

This has been a heck of a ride through the history of quantum physics, huh? We're going to finish strong with a glimpse of the state of quantum physics as we head into the future and maybe toss in one last mind-warping topic. If you guessed string theory, you'd be correct! Let's ease into that, though- we'll start with something a little lighter and walk through quantum physics in action in the things that we use every day. Then we'll attack string theory. It'll be fun!

Everyday Objects Doing Extraordinary Things

In your pocket, right now, you hold the key to all of human knowledge. It's your cell phone- and it runs on quantum physics. No, really. There are a couple of different applications of quantum mechanics and quantum chemistry going on inside your tiny personal computer, communication device, and positioning beacon. Oh, you've never thought about your phone in quite that way? It certainly makes a difference in how you consider this amazing little machine.

First, let's take into account the computer chip itself. They are a marvel of quantum technology. Most chips are made from silicon, an abundant element with some unique characteristics. When silicon is layered up in a certain way, its electrons start to create a band structure- simply put, the electron waves harmonize, and the silicon becomes able to conduct electromagnetic waves in a predictable pattern. Using band structure patterns is something quantum physicists rely on to give them feedback about how the atoms within a material are behaving because it is an observable phenomenon. It gives them the ability to determine whether a material will be a good conductor. Silicon fits the bill, and along with technological advancements, this has lent the ability to build smaller and smaller computer chips and shrink the size of computing machines from taking up whole buildings to fitting in our jeans pockets.

Your cell phone is also capable of sending and receiving data across a cellular network. What does that mean exactly? These networks rely on radio signals to allow people to send and receive calls, text messages, emails, and other communications right on your phone, anywhere there is cellular coverage. Cell towers are now in 95% of the world's inhabitable spaces, some of them "hiding" as trees, and they use radio waves of varying frequencies to relay data to one another, finally reaching your device, which is encoded with your personal frequency attached to your phone number. Pretty cool, huh? Even more remarkable when you consider that in less than four decades, the technology has become so advanced that the giant bag phones of the 1980s have become a novelty item and museum dust collectors.

Cellular towers use different frequencies to avoid jamming their own signals and also operate at varying wavelengths depending on the area in which they are built. Rural towers use lower frequencies with longer wavelengths to be able to cast their net to a larger area, and urban towers use higher wavelengths and frequencies to be able to penetrate the maze of buildings and concrete. This also leads to the need for more towers and cells in urban areas because the high-frequency waves can't travel very far. As many cellular providers upgrade their equipment to ultrafast next-generation equipment, they are also shrinking the size of the cellular transmitters to fit into urban spaces and onto utility poles. This allows them the put more cells into smaller places and not look as obvious as the old "tree" towers, which are ridiculous, anyway. Cell signals rely on 'line of sight' to be able to communicate with each other, and those towers always end up being way taller than the forest around them. But, they tried.

Another fantastic thing your cell phone can do is tap into global positioning satellites, and that's all thanks to quantum physics, too. GPS capabilities are built into most computer-driven devices these days, which is why your computer at home can show you which stores are nearby, and you can look at detailed maps of your neighborhood and pretty much anywhere on earth. GPS uses triangulation to determine your location, which is relayed from your device to satellites orbiting the Earth to find the location where you wish to go. If that isn't science-y enough, to begin with, it's the mechanism behind this action that is really dependent on quantum physics.

Every time a signal is picked up by a positioning satellite, the computer within the satellite begins performing a series of rapid calculations to convert your location from a coordinate into a length of time. Using the calculations of several satellites at once, the system can determine your location within a small radius and tell you how long it will take for you to get from Point A to Point B, which it has also calculated to within a few hundred feet. Global positioning satellites depend on atomic clocks to calculate this timing, and those clocks are run by tiny nuclear engines powered by atoms whose exact decay takes one second to release an electron. So the next time you're heading to an unfamiliar place and take your phone out of your pocket to get directions and travel times, think about all the marvelous machinery, driven by quantum physics, that is performing what seems like a simple task.

Got the World on a String

Okay, deep breath, let's tackle string theory. As we know, physicists have spent entire careers trying to develop a unified field theory, and it's always gravity that doesn't cooperate. Being about to have one giant, provable body of scientific work that explains everything in the universe would be awfully convenient, wouldn't it? Except that of all the fundamental forces of physics, we have yet to be able to observe gravity on its particle level- gravitons are still an object of mere speculation. What we do know is that wave-particle duality is a genuine, observable phenomenon, and that gives us the basis for string theory.

String theory was first introduced in the 1970s as a potential candidate for the theory of everything. It puts forth the hypothesis that perhaps particles aren't moving independently of each other but are instead the ends of invisible cosmic strings. Some particles may mark the end of open-ended strings (a particle at each end), and some may be the closing point where the two ends attach in a singularity. String theorists, and remember this is a significantly simplified explanation, believe that the open strings represent the strong, weak, and electromagnetic forces of nature, and the closed strings represent gravitational forces. By studying how these strings interact, it would be possible to trace the source of gravity and eventually find its particles. Right?

Sort of. String theory didn't initially account for the strings coming apart and joining together in different places in spacetime, and so they wanted to put them on a plane to control them. Putting them on a plane meant limiting the strings to set dimensions, and limiting the strings to set dimensions meant that string theory could no longer be used to hypothesize about controlling the fundamental forces to achieve accelerated passage through spacetime. So the next thing that string theorists did was toss in a "whatever" clause. They said that maybe there are up to three new dimensions we just haven't been able to discover yet.

String theory might seem far-fetched, and it's far from having been proven, but it does have its roots in some of the most substantial quantum physics contributions, so it shouldn't be thrown out just yet. String theory, like any other unified field theory, wants to marry relativity (dealing with the largest cosmic objects) and quantum theory (dealing with the tiniest objects) into one theory of quantum gravity. By alleging that each particle is actually one end of a cosmic string, and thinking about Bohr's theory of complementarity, then both ends of the string should balance themselves out. Some theorize that there is a fermion on one end and a boson on the other, and when they meet in a closed loop, the resulting collision releases energy and a graviton. If that could be proven to be accurate, then string theory might really have something going for it, but until then, we're stuck not being able to harness the power of gravity for our benefit.

One of the biggest proponents of string theory is physicist Edward Witten, a mathematical and theoretical physicist at Einstein's old stomping ground, the Institute for Advanced Study at Princeton. Witten is the developer of the M-theory version of string theory, and he has long proposed to solve the problem once and for all with math. Except in all his years of working on it, the math hasn't added up right yet. The amount of energy needed to produce a graviton doesn't match the amount of energy of the particles entering the collision. Mass-equivalency tells us that something is off with those calculations, and no one is quite sure how to fix it.

The exciting thing about string theory, and the one that keeps people from giving up and looking elsewhere for the theory of everything, is the potential for exploring untapped dimensions. If just by attaching one end of a cosmic string to a different particle, we could change the very fabric of spacetime and allow entrance into another dimension, then why wouldn't we be excited about string theory? That's an incredible, almost unimaginable possibility. String theory gives us a new way to think about how all matter is connected in a great cosmic dance.

String theory has its many detractors, and one person who never really bought in was Stephen Hawking, perhaps the brightest mind in theoretical physics since Einstein himself. It may be that Hawking really thought string theory was bunk, or he was concerned with pushing his own theory of everything, but Hawking was sure that string theory was a long way from being proven. Hawking was also concerned with finding a resolution to the dilemma of gravity, but he centered his work on the exploration of black holes, not necessarily as wormholes between known areas of spacetime, but as living, breathing portals into unknown dimensions. Despite his groundbreaking work on black hole radiation, gravitational fields, and the black hole information paradox (does physical data also get lost in the gravity of a black hole?), Hawking himself was never able to reconcile a unified field theory.

Perhaps there is no theory of everything, and these great scientific minds are simply chasing something that doesn't exist. One thing is for sure, though, and that is that they aren't going to stop looking. If a theory could be developed and proven that joins together quantum theory and relativity in a working, observable model of universal truth, that scientist or research team would go down in history as having a brilliance surpassing all the minds that came before. And, you also get a bunch of money when you win the Nobel Prize, so that's pretty cool.

Turning Science Fiction into Science Reality

Science fiction is one of the world's most popular genres of entertainment, and no more so than in the mid-to-late 20th century and early 21st century. Science fiction blurs the lines of reality and gives us an escape into a life of time travel, long-range space missions, cloning at the touch of a button, and all sorts of quantum physics marvels. Yes, science fiction is a little guilty of bending the laws of physics, but that's what makes it fun. The flipside of this law-bending is that real scientists are often inspired to create what they see and read in science fiction.

Remember the original Star Trek series, with their tricorder devices and flip-open communicators? In the 1960s, such technology seemed like a thing of the far future, but now we have cell phones and diagnostic scanning equipment that function in much the same way as their fictional counterparts. That's because physicists and quantum engineers were able to look at these machines and reverse-design the components that make them work in the real world, using mathematics and the knowledge of electromagnetic waves to power them. It's incredible, really.

Another science fiction element that is close to becoming a reality is teleportation, and I think we can all agree that this development is beyond exciting. In 1997, two separate groups of scientists were able to achieve quantum teleportation of photons. It may not be much, but it is a start. The theory of teleportation involves the breakdown of matter at one point, transmitting that matter's energy to another point, and then converting the energy back into matter in precisely the same way it put together, to begin with. It sounds really good in theory, but when you factor in the uncertainty principle and any other number of the laws of physics, it's really not that easy. Still, scientists are determined to unlock the mysteries of teleportation, and starting with particles seems like a logical jumping-off point. Computerization makes this task a bit easier, as it removes the human element of reassembling the matter and makes it able to be done with greater precision. It's definitely something to keep an eye on in the coming decades.

Science fiction also gives us quantum physics wonders such as food replicators, cloning devices, holograms, and long-range space travel. Let's break down how these things could become a reality, although some of them are likely not possible within our lifetimes. Food replicators, as seen on Star Trek and other science fiction programs, deliver items based on the user's specifications and in quite a short period of time. This seems a little far-fetched, as we know mass-equivalency is real, and you can't make something out of nothing.

In order for a device like a food replicator to become a reality, the machine would have to have a database containing or able to access the whole of knowledge of the atomic makeup of every single item it is being asked to create. Not only that, but it would also have to be able to harness the atomic energy needed to turn that energy into mass in the exact correct configuration so as to produce the requested item. To be honest, we wouldn't hold our breath waiting for replicators anytime soon.

Cloning devices might be less of a stretch for quantum physicists to work out. Perhaps not the human body, just yet, although geneticists and biologists are working on that, but twenty years ago, could you have imagined the 3-D printer not only being invented but being an appliance people could have in their homes? 3-D printers harness the power of quantum physics within a computer and transform a base material into the desired object, all with the touch of a button. Cloning devices may not be far behind, and physicists are hard at work around the globe trying to marry the math and the function of the technology. Again, some pretty crazy stuff.

How about holograms? They are the stuff of science fiction legend, but can modern quantum physics devise a way for people to use holograms in an everyday setting? The answer is yes, resoundingly. As we increase our understanding of the electromagnetic spectrum, we can learn to harness light waves and use computers to program them to behave exactly the way we want them to, and for holograms, that means producing and projecting a three-dimensional replica of something that's not really there. So far, physicists have been able to do this with lasers on a very small scale, but if the march of time and technology tells us anything, we know that's just the beginning. The researchers who pulled off this small-scale hologram at Brigham Young University in 2018 said it was almost like building a 3-D Etch-a-Sketch toy, but with colored lasers. Fun!

The biggest question facing quantum scientists remains that of long-range human space travel. Science fiction sends living beings all over the known and unknown universe, with seemingly little regard for gravity or spacetime. Wormholes exist as space highways, and no one ever runs out of spaceship fuel. But how can scientists get us from science fiction to science reality? Can we put humans on Mars or even further? What's the hold-up? There are several factors at play when it comes to putting people on a rocket and sending them out into deep space.

First, we cannot travel faster than the speed of light- nothing can. It takes a really long time to travel through space if you can only go so fast. We would need to develop engine technology that is safe and has a renewable energy source, as well as a spaceship that can protect the humans within from being torn apart by gravitational forces at high speeds. A person would, theoretically, die almost instantly if gravity exceeds nine times the force it is on Earth because the heart can no longer pump blood at that force. To accelerate a spaceship fast enough to achieve long-range space travel would likely cause deadly gravitational forces on the astronauts. So there's that. It looks like we aren't making the jump to hyper-speed any time in the near future. Humans are just too fragile.

That doesn't mean that physicists aren't trying to figure these things out. We have sent many a deep-space probe into the great beyond based on current technologies and a lot of mathematics. Physicists are behind the development of battery power, communications arrays, trajectory math and calculating the gravity of celestial objects, and just about every other thing that goes into designing and building the things we send into space. Just because humans can't go everywhere we'd like to yet, doesn't mean we can't send machines to do the exploratory work. It just takes a long time to get there. Until we can crack the code on moving things faster than the speed of light, it seems we will have to obey the cosmic speed limit.

As frustrating as it is for physicists not to have yet developed a comprehensive working theory of everything, they are still plugging away at it. There are other exciting developments in the field of quantum physics, too, and it's going to be a wild ride to see what the future holds. Quantum physics has officially entered a new era of technology, and this is showing up in products that we will all one day be able to have in our homes. We see it already in entertainment and computing devices. The earliest television sets made use of cathode-ray tubes that excited the particles within and arranged them in a way so as to take in the information from an electromagnetic wave, channel it through the tubes, and produce a moving picture on a screen.

Today's televisions use a smaller, more energy-efficient method of creating a picture out of electromagnetic waves, and for many devices, that's the LED or light-emitting diode. LEDs can be used in a wide variety of applications and have replaced large lightbulbs, overheating neon light tubes, and even gas-filled halogen headlight bulbs on cars. LEDs are the result of years of work by quantum scientists and engineers, and we have almost begun to take them for granted. The next generation of LED science promises to be even smaller and more energy-efficient, providing us with a whole new world of appliances and conveniences, all by harnessing the power of the visible electromagnetic spectrum.

In a previous chapter, we also talked about practical applications of lasers in medicine, but you also see lasers in many other common settings. Lasers are used to scan your groceries at the supermarket, and if you're of a certain age, you'll remember what a neat thing that was when it first became widespread. Lasers and light cells are also used in touchless technology like automatic bathroom fixtures. Lasers are also used for security purposes to determine when boundaries have been breached. So what's the future of laser technology? Lasers remain as popular a field of research as they ever have, and we're not talking about the tiny home models that you use to play with your cat. Laser is an acronym for Light Amplification of Stimulated Emission of Radiation, and the big-time laser scientists are playing with some heavy-duty stuff. These lasers are being turned into weaponry, carbon-less power systems, and systems that detect defects in combustion engines. And, as we saw in the case of researchers at BYU, lasers were used to create holograms, which is pretty awesome stuff, too.

Quantum computing is another thing that's starting to make big waves across the tech world. By learning how to shrink the size of computer chips and simultaneously increasing their computing power, the future of computing is upon us. Quantum physicists are the ones who assist computer engineers with the design of these super machines, and they are working together to build devices that can complete literally thousands of computations per second. These machines are also the next evolution of machine learning, which is a fancy way of saying that they retain and build upon the knowledge of the previous computations. Like a toddler learning to talk, artificial intelligence is improving by the day- but at the end of that day, it's important to remember that computers, lasers, robots, all of these things, are just matter being run by stimulated electrons. They cannot genuinely think on their own without input from human developers.

Separate but related to these developments are the networks that these computers will communicate with each other across. We've all heard the term "bandwidth," and that's a way of relating how much room the data has to cram into to travel. Think of it like this- the electromagnetic particles of data are vehicles trying to speed down a road. If there are 100 vehicles and only one lane open, it's going to take a long time for them all to get to their destination. But if it were a five-lane highway, you could have 2o vehicles travel per lane, and they would get there four times faster. The wider the highway, the faster the vehicles can travel, and the broader the bandwidth, the quicker the data can be transferred. Physicists and network engineers are always working in conjunction with computer scientists to ensure that their technologies can function together and complement each other's strengths.

Finally, let's take one last look longingly out into space. In 2019, scientists released the world's very first photograph of a black hole. If that isn't cool enough by itself, it's how it was taken that really captures the imagination. The Event Horizon Telescope team was made up of physicists, electrical engineers, computer scientists, astronomers, mathematicians, and support staff from around the world. These researchers collaborated to turn a global network of radio telescopes into one giant, powerful telescope by pointing their waves at the same object, in the same frequency, at the same time. While this might sound simple, the calculations involved in pulling this off were immense. The resulting data was more than enough to produce a clear photograph of a supermassive black hole in the Virgo galaxy, more than 54 million light-years from Earth.

The accomplishment is really a good metaphor for the whole of quantum physics. We've seen that each discovery builds upon the one before it. We've learned how cooperation among scientists leads to bigger and better things. We've also seen how quantum physics encompasses so many different sub-disciplines. From mechanics to chemistry, astrophysics, and quantum mathematics, we know that when people who specialize in the various areas of quantum physics come together in the name of science, great things are possible. When the atom was first described by Democritus back in 400 B.C.E., it's unlikely that he ever dreamed how far his ideas would go. Maybe that's really the heart of quantum physics- it is a field built on the backs of scientists who dared to dream about the infinitesimal and the infinite and then had the courage and the audacity to try to figure out if they were right or wrong.

And now we've come to the end of our time together. However, the journey of quantum physics is one that has no foreseeable end. We hope you've learned enough in this primer to pique your interest in learning more. There are always new and exciting things to be discovered when it comes to the universe's unseen forces and the building blocks of matter. Thanks for sticking out the whole book with us- we hope you enjoyed it all in the name of science!